The Unity of Worlds and of Nature

Three Essays on the Spirit of Inductive Philosophy; the Plurality of Worlds; and the Philosophy of Creation

BADEN POWELL

CAMBRIDGE
UNIVERSITY PRESS

CAMBRIDGE UNIVERSITY PRESS

Cambridge, New York, Melbourne, Madrid, Cape Town, Singapore,
São Paolo, Delhi, Dubai, Tokyo

Published in the United States of America by Cambridge University Press, New York

www.cambridge.org
Information on this title: www.cambridge.org/9781108004565

© in this compilation Cambridge University Press 2009

This edition first published 1856
This digitally printed version 2009

ISBN 978-1-108-00456-5 Paperback

CAMBRIDGE LIBRARY COLLECTION
Books of enduring scholarly value

Religion

For centuries, scripture and theology were the focus of prodigious amounts of scholarship and publishing, dominated in the English-speaking world by the work of Protestant Christians. Enlightenment philosophy and science, anthropology, ethnology and the colonial experience all brought new perspectives, lively debates and heated controversies to the study of religion and its role in the world, many of which continue to this day. This series explores the editing and interpretation of religious texts, the history of religious ideas and institutions, and not least the encounter between religion and science.

The Unity of Worlds and of Nature

Baden Powell (1796–1860) was a mathematician who held the Savilian Chair of Geometry at Oxford, and was also a priest in the Church of England. He was a defender of the claims of new scientific discoveries in the face of Christian orthodoxy well before Darwin published the theory of evolution, and drew a clear distinction in his thinking and writing between moral and physical phenomena, as being independent of each other and the fields of completely different study. Darwin himself wrote, in the 'Historical Sketch' at the beginning of the third edition of On the Origin of Species, 'The "Philosophy of Creation" has been treated in a masterly manner by the Rev. Baden Powell, in his Essays on the Unity of Worlds, 1855. Nothing can be more striking than the manner in which he shows that the introduction of new species is "a regular, not a casual phenomenon". The second, enlarged and revised edition of 1856 is reissued here.

Cambridge University Press has long been a pioneer in the reissuing of out-of-print titles from its own backlist, producing digital reprints of books that are still sought after by scholars and students but could not be reprinted economically using traditional technology. The Cambridge Library Collection extends this activity to a wider range of books which are still of importance to researchers and professionals, either for the source material they contain, or as landmarks in the history of their academic discipline.

Drawing from the world-renowned collections in the Cambridge University Library, and guided by the advice of experts in each subject area, Cambridge University Press is using state-of-the-art scanning machines in its own Printing House to capture the content of each book selected for inclusion. The files are processed to give a consistently clear, crisp image, and the books finished to the high quality standard for which the Press is recognised around the world. The latest print-on-demand technology ensures that the books will remain available indefinitely, and that orders for single or multiple copies can quickly be supplied.

The Cambridge Library Collection will bring back to life books of enduring scholarly value (including out-of-copyright works originally issued by other publishers) across a wide range of disciplines in the humanities and social sciences and in science and technology.

THE

UNITY OF WORLDS

AND OF

NATURE:

THREE ESSAYS

ON

THE SPIRIT OF THE INDUCTIVE PHILOSOPHY;
THE PLURALITY OF WORLDS;
AND THE PHILOSOPHY OF CREATION.

BY

THE REV. BADEN POWELL,

M.A. F.R.S. F.R.A.S. F.G.S.

SAVILIAN PROFESSOR OF GEOMETRY IN THE UNIVERSITY OF OXFORD.

SECOND EDITION,

REVISED AND ENLARGED.

LONDON:

LONGMAN, BROWN, GREEN, LONGMANS, & ROBERTS.

1856.

The right of translation is reserved.

PREFACE

TO

THE SECOND EDITION.

A SECOND edition having been called for, I have endeavoured to render the present volume less unworthy
of the public attention by a careful revision and cor-
rection of the text, and especially by availing myself
of the valuable suggestions of several friends, as well
as of those criticisms in periodicals which, from bearing any evidence of honesty, fairness, or ability,
seemed deserving of notice. I have thus been led to
make many additions, besides numerous lesser alterations; — but, to no modification of the essential argument, which it is hoped those changes will only
render more clear and forcible.

My Third Essay more especially, having been
publicly adverted to by Mr. W. J. Hamilton, P.G.S.

in his anniversary address to the Geological Society,
1856, I have felt it necessary to revise carefully the
points in which he has criticised my argument,
(though in no instance impugning my facts,) while he
has given a flattering general commendation of the
object and tenor of my work.

At the same time, I cannot help remarking as
curious, how eagerly eminent geologists seem to single
out, for the display of their controversial zeal, the obnoxious topic of the Development hypothesis, which,
after all, I do *not* maintain.

With regard to some points more properly of a
theological kind, I could have wished to add further
elucidations of them, especially as in some quarters
my meaning has been considered ambiguous, and
perhaps some misapprehensions entertained with
respect to it; but finding it impossible within the
necessary limits to discuss such points as they deserve, I have contented myself for the present with a
few verbal corrections to render the meaning clearer,
in the hope, at a future time, of going into such
discussion in another series of essays.

In the original Appendix (No. I.), I had annexed
some elucidations of the logical principles of the argu-

ment especially as bearing on the theory of Induction. In this edition, I have added some further remarks in order to point out the connexion between the views here adopted and those of Kant and some other metaphysicians.

One or two other additional illustrations of different parts of the argument are also annexed.

PREFACE

THE FIRST EDITION.

THE three following essays, though somewhat connected in subject, are yet each distinct and complete in themselves, having been originally composed at different times and with separate objects. Hence there will probably be found in some parts *repetitions :* but on the whole it appeared preferable to allow these to remain, rather than by omissions and alterations to render less complete and continuous the argument of each essay in itself. And the few topics which belong to them in common will, in most. cases, be found treated under somewhat different aspects, according as the particular argument in. each instance required.

The First Essay consists mainly of an amplification of a few paragraphs in my paper " on Necessary

and Contingent Truth" in the Oxford Ashmolean
Memoirs, 1849, in reference to which I felt it
desirable to explain and illustrate more fully some
points there but imperfectly treated; as well as some
other topics related to them, and which have of late
years been the subject of considerable discussion:
some of which were also considered in my work on
" the Connexion of Natural and Divine Truth,"
1838. More precisely, the subjects of the primary
grounds of inductive reasoning, and the theory of
Causation, have long since appeared to me to be commonly involved in much confusion of thought, which
has, as I think, been rather increased than diminished
by some recent discussions from which we might have
hoped for greater enlightenment; — and which appears to me to be the source of many unhappy difficulties and objections connected with the so-called
doctrine of " final causes," and the evidences of natural theology generally.

To the object of clearing up some of these difficulties, and inculcating better views, some parts of
my former work last referred to were devoted: And
to the argument there pursued (so far as I am aware)
no substantial objections have been alleged. Yet the

frequent reproduction of the same original confusion of language and thought, in otherwise able and valuable writings at the present day, renders it not useless to recall attention to some of those considerations by which, I believe, the whole subject is put on a more satisfactory and unobjectionable basis.

Many of these topics, it will be evident at first sight, are coextensive with those so elaborately and profoundly treated in Dr. Whewell's Philosophy of the Inductive Sciences and in Mr. Mill's Logic. If I have made very few specific references to either of those treatises, it has arisen from no want of respect or consideration for either of the distinguished authors; but rather from an opposite feeling of high general esteem for the ability with which they have treated the subject, I entertained an unwillingness to appear to enter into direct controversy, in some material questions on which I have been constrained to hold opinions somewhat differing from those of both writers, though, in general, more nearly coinciding with the latter.

If the grounds on which I maintain my views shall be found sufficiently indicated and explained, I trust the candid reader will be as well prepared to come to an unbiassed opinion on the points in question as if they were urged with a greater degree of critical detail; and the opinions which I controvert will be equally marked out, without more minute reference to the particular authors.

The Second Essay was called forth by a perusal of the two able and interesting works on the question of the Plurality of Worlds, which have of late attracted such an unexpected degree of public attention; an interest which, even up to the moment of bringing out this volume, does not appear to have abated, if we may judge from the numerous other publications since announced on the same question.

With respect to the author of the " Essay on the Plurality of Worlds," while it would be absurd to pretend ignorance of his real eminence, I have throughout felt it would at the same time be improper to refer to his opinions, otherwise than as those sustained by the masked character under which, doubtless, for the greater freedom of such discussion, he has thought fit to veil academical dignity.

The controversy itself, as to the question of inhabited worlds, appears to me of comparatively little moment: it is rather for the sake of more general considerations involved, that I have been led to enter into the discussion, and, in some measure, to hold the balance between the two disputants. Those broader principles are closely connected with the subject of the First Essay.

The collateral questions introduced into the Second Essay have also an immediate bearing on the subject of the Third. The inquiry into the *present* condition of planetary worlds is closely connected with that of their *past* state and probable origin; and this with the general question of the history of *creation*, so far as it can be traced on physical grounds. But this subject again, is one which has of late years extensively occupied the public attention; especially from the extraordinary popularity attained by the " Vestiges of the Natural History of Creation," and the controversies to which that work has given rise. In those controversial discussions, it cannot but be matter of regret that so acrimonious a tone, little suited to eliciting the truth, should have been adopted by some of the writers. Hence it seemed to me that a more calm and philosophical analysis of the whole question was much needed; and in some measure to supply such a review of the general principles and grounds on which all speculations of the kind should be conducted, as well as to examine dispassionately into the alleged religious bearings of any theories by which some part of the steps and processes of creation might be explained, has been the aim of the Third Essay.

It should perhaps be observed that if, in those passages where I have spoken of the evidences of natural theology, I have professedly restricted my remarks to the *physical* portion of the argument,— it is not from at all disparaging or overlooking the *moral* and *metaphysical* portions, that I have not adverted to them, but solely because they are not immediately connected with the more direct object of these Essays.

A similar remark ought, also, to be made with respect to the very brief and inadequate mention made of some other points of deeper import to the belief in revelation ; to which I could willingly have devoted a more extended discussion than it was possible within my present limits to give them.

ANALYTICAL TABLE

OF

CONTENTS.

§ v. — FINAL CAUSES, AND NATURAL THEOLOGY.

ESSAY II. — ON THE UNITY OR PLURALITY OF WORLDS.

§ I. THE ARGUMENT CONSIDERED IN A PHYSICAL AND PHILOSOPHICAL POINT OF VIEW.

a

APPENDIX.

APPENDIX

ESSAY I.

ON THE SPIRIT OF
THE INDUCTIVE PHILOSOPHY.

§ I. — THE INDUCTIVE PRINCIPLE.

GALILEO, 1590.

"Opinionum commenta delet dies,
Naturæ judicia confirmat." — CIC.

THE characteristic nature, genius, and grounds of Introductory remarks. the inductive philosophy have been much discussed of late years, and under considerable varieties of aspect, by different parties. Whilst some have carried out their view of its principles into metaphysical abstractions often hardly intelligible, others have sought to narrow them to the results of mere sensible experience; and whilst the one would connect its aims with a higher intellectual philosophy, even verging on the mystical, the other school would lower its objects to the mere empire over matter, and the attainment of utilitarian ends.

Distinction of sensations and ideas.

More precisely, an inquiry into the essential grounds and principles of induction involves the general question of what has been termed " the fundamental antithesis "* of sensations and ideas, facts and theories; in a word, of two essentially distinct and independent sources of all knowledge, the external and the internal — observation by the senses, and ideas originating in the mind itself; while it is only by the application of the latter to reduce to system the materials supplied by the former, that any real philosophical theory can be constructed, the crude results of observation be converted into an inductive theory, or sense elaborated into science.

Thus ideal conceptions, the pure offspring of mind, the mere creatures of intellect, seem to exercise a sort of plastic power over the mass of

---

* See Dr. Whewell's two able memoirs " On the Fundamental Antithesis of Philosophy."   Cambridge Phil. Society Transactions, 1848.

material results, giving them a fresh character and scientific significance; and thus we are enabled to make that ascent from facts to laws, from laws to causes, which is the aim and boast of the inductive philosophy. Such views, carried out in some instances to speculations of a kind still more remote and hardly comprehensible, have been adopted by many at the present day: while, on the other hand, the "positive philosophy" is characterised by a tendency to the contrary extreme of discarding all reference to those higher intellectual principles, reducing all science to the naked results of observation and calculation, and all idea of causation to that of mere invariable sequence of phenomena.

In looking more precisely to the meaning of the term *experience*, if we understand it literally as the mere collection of facts, such as sense and observation *directly* furnish, and the rejection of everything which is not, in this restricted sense, properly learnt by it, then, indeed, there is an end put to all really scientific or philosophical investigation; and beyond the narrow circle of those facts we can never enlarge our conceptions or raise our contemplations.

The slightest consideration, however, will show

Meaning of the term "experience."

that the term *experience,* even in the simplest case, must be understood in a wider sense : whilst the logical analysis* of induction exhibits a syllogism in which a large assumption is necessarily implied, beyond and independent of any accumulation of facts.   Thus every induction is seen essentially to involve a *certain amount of hypothesis,* — a certain assumption of more than the bare facts themselves seem strictly to warrant.   We form intellectual conceptions of a nature more general than the mere enumeration of a number of instances, how-ever many ; and thus supply " the string on which " (as Dr. Whewell happily expresses it) " the pearls are hung ; " and perceive, according to the illus-tration of another able writer†, how " philosophy proceeds upon a system of credit, and that, if she never advanced beyond her tangible capital, her wealth would not be so enormous as it is."

It is certainly not the mere *number* of instances which constitutes the strength of an inductive con-clusion ; but it is the *kind* and *quality* of them, as

---

* See Archbishop Whately's Logic, book iv. ch. i. §§ 1, 2.
† Outlines of the Laws of Thought (p. 312.), by Rev. W. Thomson, M. A.: London, 1849.

bearing on the manifestation of the existence of certain relations among them, connecting them together by *analogy*. If the individual facts be thus connected, or of the *right sort*, a comparatively *small number* of them will be convincing, when in other cases the most laborious accumulation will be fruitless and unsatisfactory, as wanting in a real connection of analogy. When, however, that essential condition is secured, it then infallibly happens (as has been well said) that a " vague and local idea . . . . passes through the mint of a very few decisive experiments into the treasury of accepted truths."*

In arriving at any general inductive conclusion, then, *something* is clearly superadded to the mere mass of facts; the question, is *what* is it? In the simplest case, that of knowledge acquired by the senses, something more than mere sensation is implied: besides sensations conveyed *to* the mind, there must be corresponding ideas excited or formed *in* it. All observation which involves *mind* involves *theory*: the *facts* of sense must be *idealised*. Of

What it is which is superadded to sensible facts.

---

* Rev. W. V. Harcourt's Letter, &c., Phil. Mag. 1846, p. 76.

the truth of this no reflecting person entertains a doubt: the sole question is, *how* it is effected, and *whence* these ideas are formed.

Supposed
inherent
faculty or
principle. According to one school, these phenomena are referred to a peculiar principle, supposed to be implanted in the mind, not to be further analysed: a special faculty, producing a distinct mode of conviction; a kind of assurance, prior to and independent of external sense, and derived from the interior resources of reason; an inherent intellectual element, which warrants us in extending our conclusions beyond the mere limits of observation, and in inferring intuitively and certainly the future or unknown from the past or known. Or, more precisely, certain fundamental conceptions are supposed primarily and originally formed within the mind itself, derived somehow from its interior resources, without any reference to external sensation; and the introduction of these conceptions (differently modified according to the nature of the respective subjects) impresses the proper form on the collected facts. And it is from the fundamental ideas thus entering into combination that the attributes of universality and necessity are acquired

by our conclusions and belief, and a certainly attained
on *à priori* grounds which no mere observation could
furnish.

Another school, discarding all reference to any
intuitive or internally created ideas, analyses the
intellectual process into its elements, and shows
that through successive steps of abstraction, from
the simple collection of facts, we advance to theories
which are true just in proportion as we are guided
by the right perception of analogy and the important rule of correcting one generalisation by
another, and thus, that all knowledge is ultimately
derived from observation.

Another view. Gradual process of abstraction and gene- ralisation.

The theory of intuitive or internal principles
undoubtedly appeals powerfully to the imagination.
Nothing seems more natural or plausible than to
refer everything to *ultimate principles* originating
in the mind : it saves the labour of further analysis,
and supplies a specious explanation of intellectual
phenomena, which seems to gratify at once the
desire of penetrating the secrets of our nature and
the love of the mysterious, in appealing to great
but hidden causes within us : a species of occult
philosophy, which seems eminently to harmonise

Idea of in- tuitive principles natural.

with the mysticising tendencies of the age; but
which, nevertheless, appears to be conceived in a
spirit very opposite to that of the simple and
positive character of the inductive method, and,
though sanctioned by great names, seems rather to
be a retrograde movement, and to evince a lingering
attachment to the scholastic mysticism, or to be in
some sense a revival of it.

Ought to be
analysed up
to simpler
elements.

That we are *naturally prone* to entertain such
notions may be very true; yet it may happen in
this, as in many other instances of what we are
prone to do, that we do *wrong*. But the more
strict metaphysical inquirer will acknowledge that
it is unphilosophical to imagine peculiar and unknown mental principles, if processes carried on
through already acknowledged intellectual powers
can be shown to suffice for explaining the facts.

In the present case, indeed, as in other inquiries, it may be perfectly allowable in the first
instance to set down any outstanding class of phenomena as *provisionally* something *sui generis*, and
of an elementary character, just as in chemistry
we may regard any new substance as *elementary*
while it is as yet *undecomposed*; but still it is the

aim of the chemist to decompose it if he can. In
the same way there may be a multitude of ideas,
impressions, intellectual sensations, and the like,
which may at first seem like elementary principles;
but which, nevertheless, it should be the aim of
the metaphysical analyst to reduce into their component simpler elements if possible.

In such cases, the powers of imagination may be
appealed to; and doubtless those powers are sufficiently prolific in suggesting theories. The minds
of the ancient philosophers teemed with speculative
schemes of nature, before any study of facts had
furnished them with substantial materials. Humboldt has well observed, that "long before the discovery of the New World it was thought land could
be seen in the west from the Canaries and the
Azores. They were phantasms not produced by any
extraordinary refraction of the rays of light, but
merely by a longing for the distant, for that which
lies beyond the present. The natural philosophy of
the Greeks, and the physics of the middle ages and
even of much later centuries, presented swarms of
such fantastic forms to the imagination. The mental
eye still essays to pass the horizon of limited know-

Power of
imagina-
tion.

ledge even as the material eye endeavours to pierce
the natural horizon from an island height or shore.
Faith in the unusual and wonderful gives definite
outlines to every product of imagination; and the
realm of fancy, a strange land of cosmological,
geognostial, and imaginative dreams, is incessantly
blended with the world of reality."*

Yet mere imagination, however powerful and
prolific, will avail little for creating any theories
which will stand the test of observation, or which
have any real application in nature.

Something more than imagination.

But, from considering the nature of our generalisations, it is argued, that we must necessarily
obtain ideas from *some other source* than *sense,* or
that the mind possesses a peculiar power or faculty
of acquiring a higher degree of certainty from
within than experience can give from *without*. Or,
again, it is said, in such cases as mathematical
theorems, the mind attains certainty quite independently of experience; whilst in other cases,
such as limited inductions in subjects little known,
it has no certainty beyond the mere facts which are

* Cosmos, p. 84., 1st trans.

directly presented to it. Why, then, is the mind
so confident in one case and so cautious in the other,
unless there be a real difference in the faculties
brought into play in the two respective cases ?

When we analyse the process logically, it is
manifest that, in induction, what is superadded to
a mere collection of facts consists precisely in the
assumption *" that all phenomena of the kind in question are similar to the few actually examined."* *

Logical
analysis of
induction.

This, abstractedly speaking, is perfectly general,
applying to all cases of induction alike. But we
may here notice an important distinction which has
been drawn between two kinds or classes of induction : (1) that by experiment, (2) by observation.

In the first, the assumption just mentioned is at
once warranted from the circumstance, that here the
reasoner himself constitutes, by selection, the class
of objects to which his conclusions refer; this process
therefore involves directly the truth of the assertion,
" all objects of the class examined are like this ; " as
for example, a chemical analysis of a single drop of
pure water is true for all the water in the world;
and the like.

---

* See Appendix, No. I.

In the second class, this condition does not obtain: here, therefore, further consideration is necessary. Taking this then as the more general case, I proceed to the examination of it. *

Origin of this assumption.

The question, then, is reduced to this, How does the mind come to make this universal assumption, and to be so firmly convinced of its truth?

Mental processes carried on unconsciously.

In the first place, I think it will be allowed, on reflection, that general conceptions of this kind, however apparently abstract in their nature, may be created in our minds by very simple causes, of whose operation we may yet be quite unconscious. There is nothing of which we are less conscious than the acquisition of the commonest ideas by daily experience, and the successive and gradual generalisation of that experience by the process of *abstraction*; and in this way we constantly

---

* The distinction here introduced has been acutely pointed out and illustrated by Dr. Mayo, in his " Outlines of Medical Proof, &c.," p. 4. 2nd edition, and by Mr. Mill, Logic, book iii. ch. 7. The former remarks that in the first case, " a confident assumption is obtained by a single case with only such repetition as is necessary to insure the correctness of the analysis. In the second no such confidence is obtained, the process not being exhaustive. The one seems a disintegration of facts, the other a collection of them." I venture to quote these words from a letter addressed to me by Dr. Mayo on this subject. See also Sir J. Herschel, Discourse on the Study of Natural Philosophy, § 67., and some able remarks in a recently published work, " A Delineation of the Principles of Reasoning," by the Rev. R. B. Kidd, London, 1856, p. 277.

obtain (without being aware of it) numberless pre-possessions and convictions far stronger than any systematic demonstrations can supply.

The primary assumption involved in all induction is the *presumed* uniformity of phenomena, or the conformity of other facts of the same class with that under examination to the same law or type.

Assumption of the uniformity of nature.

It is, then, perfectly true that no inductive process can advance without *the assumption of this generalising principle, which is, nevertheless, antecedent to the particular class of experimental testimonies* IN THAT INSTANCE *appealed to.* But what I would particularly dwell upon is, that *it is not antecedent to* ALL *experience;* it is some principle already established in the mind by previous abstractions, *remotely* derived from previous experience, and *specially extended by* ANALOGY beyond the precise limits of actual observation in this instance.

Idea of generalisation derived from gradual experience.

It is true that there exists in the human mind a strong natural propensity to draw *hasty inferences*, to generalise too rapidly, and to deceive ourselves by erecting conclusions on very unsubstantial and

Proneness to hasty generalisation.

insufficient data; and this is closely associated with
the fondness for tracing resemblances; being pleased
with uniformity and the contemplation of analogy,
real or imagined, where there are often but slight
indications of it, or even where the appearances of
it are in reality altogether fallacious.

These propensities are evinced more or less strongly
in different minds in the earliest exercise of their
powers: and though in matters of common life and
every-day occurrence they are soon and effectually
subjected to the corrective process of enlarging experience and reflection, which the pressing necessities
of daily existence force upon us; yet in other subjects, such as those of abstract speculation or philosophical inquiry, it may be long before they receive
so salutary a check, or at least before they come to
be really well regulated by rational principles.

Generalisa-
tions imper-
fect at first,
corrected
by increas-
ing ex-
perience.

Our FIRST *inductions are* ALWAYS IMPERFECT AND
INCONCLUSIVE; we advance towards real evidence
by successive approximations; and accordingly we
find false generalisation the besetting error of most
first attempts at scientific research. The faculty to
generalise accurately and philosophically requires
large caution and long training; and is not fully at-

tained, especially in reference to more general views, even by some who may properly claim the title of very accurate scientific observers in a more limited field.   It is an intellectual habit which acquires immense and accumulating force from the contemplation of wider analogies ; and in any one case our conviction of inductive truth is largely built up on past trial of its soundness in other cases ; and from the perpetual multiplication of such cases it obtains a perpetually progressive character of greater certainty, increasing in a rapidly accelerated ratio as experience enlarges.

By trial of theoretical suggestions in succession, and only after repeated failure, we learn their erroneous nature.   But thus by acquiring more caution and confidence and adopting better conjectures, we revise and amend our attempts, and learn to proceed on more sound principles, until we gain a habit of generalisation worthy the name of inductive power.

Again, the tendency to make the primary inductive assumption, and the extent to which it reaches, admit of many degrees.   It is found in its higher perfection in those comprehensive views which constitute the discoveries of the greatest philosophers, and in varied inferior degrees in other instances.

Tendency evinced in different degrees.

C

Feeble in
earlier
stages.

In the order of time, also, it is always evinced with far less effect in the earlier stages of scientific development, and with more full and perfect force in its later progress; whether in the infancy of science, or of the experience of an individual.

Gains
strength by
advance of
experience.

But as the cultivation of inquiry advances, the inductive process by habitual exercise derives force so naturally and insensibly, that the mind is utterly unconscious of its acquirement; and hence it is that we readily give way to the very natural, but mistaken persuasion, that the generalised idea is something inherent, or created out of the intrinsic powers of reason itself.

And from
absence of
contradic-
tory cases.

And in any case even of the most limited induction, there is one argument on which, more than any other, we always fall back with perfect confidence, and which really constitutes the main force of the evidence, viz. the assurance that *if there be any fatal exception* to the law or truth supposed to be established, *it will soon be sure to manifest itself.* The *non*-occurrence of such an exception against a supposed law is a far stronger argument than the accumulation of hundreds of instances in its favour : and this consideration probably operates far more strongly with most minds than any abstract principle of conviction.

If there be any force in what has been advanced, then, instead of any primary or inherent principle,— any original element of the mind, enabling it to see the outward world blindfold, — any intuitive internal power to create external facts, any authority derived solely from the interior resources of pure reason to show us physical and material things without reference to the senses, or the like,— the simple analysis of the case would lead us to the more sober belief that the source of inductive certainty, that certainty beyond the mere limits of sense, that superstructure larger than any foundation of facts, is accounted for by natural and acknowledged processes.

No intuitive perception of external truth.

The case resolvable into the power of abstraction.

It arises in the first instance out of the power of *abstraction*, acting with unconscious force and powerful rapidity, by whose aid the mind creates what are indeed new conceptions, yet formed only out of materials already furnished, and this *not by addition, but by subtraction* of properties and particulars.

Above all, the process derives its whole force from the discovery and acceptance of sound and well-framed *analogies*, or, as I have elsewhere said, THE SOUL OF INDUCTION IS ANALOGY; and higher, more effica-

Mainly aided by perception of analogy.

cious, and more enduring, as the analogies adopted
are more strictly accordant with the real harmonies
of nature.

Application
of mathe-
matical
reasoning.
The application of a higher reasoning to the mere
facts of observation which essentially constitutes
science throughout a large extent of physical research, is mainly effected by the application of those
systems of abstract and necessary mathematical truth
which have been independently deduced from abstractions respecting quantity in its several species
(themselves derived not less originally from ex-
periences of sensible extension, division, and nume-
ration), whence spring quantitative laws and mathematical theories, which confer on the inductive
results, whenever they can be applied, a character
of increasing certainty and power arising from the
higher capacity for generalisation. Thus the two
systems react on each other, and we are often enabled to carry on our views, and predict results to
which no mere extension of observation could have
conducted us.

The process of inductive generalisation indeed
becomes infinitely more rapid, decisive, and well-
founded, when pursued in connexion with the de-

ductive method. The application of mathematical formulas, if found to apply to the subject, not only leads with greater readiness to general laws, but carries with it a powerful presumption in many cases that it is really the exponent of some actual and higher natural analogy which we could never have collected from any mere observation of facts.

Such instances are, indeed, constantly occurring in various degrees; but, in some particularly striking cases, have evinced, to a singular extent, the correspondence between the real, but as yet unknown, laws of nature, and the abstract creations of mathematical conception: as in the well-known instances of the change of plane into circular polarisation predicted by Fresnel from the mere interpretation of an algebraic symbol, and the fact of conical and cylindrical refraction anticipated from the mathematical theory by Sir William Hamilton.

Correspondence of mathematical and physical truths.

But this assertion of *à priori* evidence is sometimes made with reference to the primary principles of all natural philosophy — the laws of motion and of equilibrium — whether in solids or fluids. It is alleged, that what is announced as the first law of motion, though it may be attested by constant ex-

Asserted à priori evidence of physical truths, examined in several instances. Laws of motion.

perience, has yet in itself evidence arising out of the nature of the case beyond all experience.

Now, in the first place, I would observe, that the very notions of a body in uniform rectilinear motion, or of forces acting on it, are essentially ideas of experience, and certainly could have no application without reference to the real existence of matter and force.

It may be maintained that the law of inertia — that a body will retain motion communicated to it *after* the direct impulse has ceased — is deducible as a consequence from higher first principles; but still those principles are themselves nothing else than more simple facts, or properties of matter, derived from experience.

It is sometimes alleged that, to maintain that a body, left to itself, will go on in uniform rectilinear motion *for ever*, is presumptuously to assert what no *experience* can ever justify ; and, therefore, if admitted at all, can only be received as an intellectual truth derived from *à priori* principles. But such perplexity would be removed if we only put the proposition thus : a body in motion, &c., *must* EITHER go on for ever, OR, its motion must be changed or stopped; but whatever changes, stops, or retards it, is a new force acting upon it, and the question is then reduced to an examination of the action of that force.

Again, it has been sometimes asserted, that the first principle of equilibrium — the foundation of the doctrine of the lever — is *axiomatic* or self-evident. Equilibrium.

Yet, without going further, it is obvious that the very idea must imply at least the existence of matter, capable of being acted upon by such a force as gravity through the intervention of something material corresponding to the inflexible straight line of theory; — ideas which can only have been obtained ultimately from experience. When some such principles have been adopted, we can then, and then only, by strict deductive reasoning from them, arrive at the theorem of the lever, which we find confirmed by experiment.*

In like manner, it has been maintained that the first principle of all hydrostatics, *the equal pressure of fluids,* is not derived from experience, but that the mind can pronounce on its *à priori* certainty. Equal pressure of fluids.

---

* See my "Essay on the Laws of Motion," and "Essay on Necessary and Contingent Truth;" Ashmolean Memoirs: Oxford, 1837, 1849.

Undoubtedly the mind can infer deductively this
great law of fluids, as a *necessary consequence from
certain other assumptions,* that is, when certain yet
more elementary properties of fluids are known,
and taken as the basis of the science; but not otherwise.

A conse-
quence
from the
nature of
fluids.

The ulterior principles to which the nature of fluids
may be reduced, may have been differently viewed
and traced upwards to more or less simple elements
by different philosophers, but all have adopted, and
must adopt, at the outset, *some primary physical* fact
or property to start from. The more simple and
general the property referred to, the more satisfactory
and complete is the reasoning; and it is the main
point in such an enquiry to determine *what* are the
fewest and *simplest* principles we can assume, in
proving these first properties and laws. Still, the
ultimate principle, however simple, and however far
back it may be traced, can of necessity be nothing
else than some *physical* fact, the result of universal
observation; such as must be even the very *existence*
of fluids, and without which no reasoning of the kind
could be applicable.

Abstract

It is, indeed, quite conceivable that a reasoning

being, who had never seen a fluid, might imagine and create theoretically the conception of such a substance, and might reason mathematically on its properties, such as would follow by strict deduction from the constitution thus assigned to it; but this would not apply to anything in nature until it were shown by experience that these properties were really manifested in some substance to which the theoretical notion might be referred.

theory may be conceived.

But physical observation necessary for application.

This is no imaginary case: it actually occurs in the speculations pursued by so many philosophers on an imagined æthereal medium. From the assumed nature of such a purely hypothetical medium, a supposed assemblage of imaginary molecules, acted on by attractive and repulsive forces and liable to agitations from without, by mathematical reasoning the whole of the refined and complicated theory of undulations has been deduced ; which, so far, might for ever remain a barren but most beautiful mathematical creation. Independent observation gives us no evidence of the existence of such a medium, and the theory is in no way founded on experience.

Example of the undulatory theory.

When, however, by the aid of the *eye*, the phenomena of optics present themselves, we find a vast

Unapplied till optical facts are introduced.

range of such phenomena which admit of a complete explanation on the assumption of this hypothesis : here, for the first time in the inquiry, a reference to anything experimental or sensible comes in. That it must come in somewhere is clear; yet it would be absurd and untrue to say that such theoretical reasoning alone can give any *à priori* certainty to the optical facts or laws to which it is applied, which must after all have been first founded on some small basis of observation.   Nevertheless, such applications of mathematics confer *the highest presumption*, little or at all short of certainty, for generalising conclusions actually observed to be true only in one or two instances.

Inverse square of the distance.

To take, perhaps, the strongest instance which has been adduced.   The law of force or intensity varying as the *inverse square of the distances*, it is alleged, and doubtless with truth, is a conception of pure reason (so far as any mathematical conception is so) from abstract geometrical considerations, which must hold good in any kind of supposed *emanation*, radiating equally in all directions from a centre, and undergoing no change of condition excepting that due to distance only.

But though these geometrical ideas throughout may be pure creations of the mind, yet the idea of any such emanation of actual force, however abstract, must have been derived from some ideas of experience, and certainly can apply to nothing in nature without reference to such sensible ideas.

Again : to take what is almost an equally striking instance, — the law of equal areas.

Equal areas proved abstractedly.

It is undeniably a pure result of reason that a metaphysical point revolving about another metaphysical point by virtue of an impulse conspiring with a centripetal force tending to that point, varying according to any law whatsoever, must describe areas proportional to the times.

But how do we get the idea of a centripetal, or of an impulsive force, unless, at least in the first instance, by abstraction from observed facts ? Wherever these forces exist in nature, we reason deductively to the conclusion of a description of equal areas, and we find it confirmed by observation.

Inapplicable without some ideas from experience.

But this is a very different thing from gaining independent *à priori* evidence for physical facts.

From expressions sometimes used, it would even seem that additional force is supposed to be given

Paradoxes.

to the argument for abstract conception as the ground
of physical truth, by the allegation, that some of
those primary abstract physical principles to which
we have referred are even *opposed* to what mere
sense and experience would naturally expect, and
must therefore be ascribed to a higher faculty of
internal reason; and this, it is also alleged, is practically evinced by the circumstance that such truths
are appropriately termed *paradoxes;* as, *e. g.*, the
primary property of fluids has led to what is called
the " hydrostatic paradox."

All truths
paradoxes
to prepos-
sessed
minds.

But this is not owing to anything in the abstract
nature of the reasoning. For what does a *paradox*
really imply? Any new truth, even a mere matter
of observation, is a paradox in popular estimation, if
it contradict a received prejudice. The existence
of Jupiter's satellites, and the fall of unequal weights
in the same time, were paradoxes when announced
by Galileo* to the Aristotelians of his day. Yet
these were facts of *observation*.

The Aristotelians had held that motion can only
be caused by something in *contact* with the body

---

* Vignette at the beginning.

moved; hence the law of inertia was a *paradox* when first asserted; and so, indeed, it continued to be long afterwards even to the Copernicans, as appears from the difficulty they felt in accounting for the continual keeping up of the planetary motions. The application of abstract reasoning in such cases tends, in fact, to *remove* and *explain* the paradox, not to create it. The startling nature of the assertion, therefore, is no proof of its being derived from any intuition superior to sense.

The question between the *inductive* and the *deductive* process is merely a question of *degree:* in some cases the abstract part of the process may be longer, and its origin more remote from material facts — in others less so.   The very same conclusion may often be arrived at by several distinct trains of reasoning, setting out from principles of lower or of higher degrees of abstraction; but there must always be, *somewhere* in the process, a recurrence to sensible experience.

Deductive proof only from physical principles more or less remote.

For instance, without any knowledge of mathematical theories, we might discover *experimentally* and empirically the laws of the motion of the *pen-*

Example. Theory of pendulum.

dulum; and so might regard them as _mere facts of observation._

But, again : if we knew in the first instance, by experimental trial, the law of _falling bodies,_ we could deduce mathematically what _must_ be the law of the pendulum, — that is, it is a necessary consequence in reason from a simpler mechanical truth, provided that reason be first furnished with that simpler truth.

But, once more : the law of _falling bodies_ itself is a necessary consequence of still simpler principles : if we knew, experimentally, the nature of terrestrial gravitation, we might deduce, by pure reasoning, the law that the spaces described under its influence by bodies falling near the surface of the earth must be proportional to the _squares of the times ;_ and thence deduce the laws of the pendulum.

But even, still further : if we investigated, _on pure theory,_ the effects of _a constant force,_ we should deduce the same law for bodies moving from a state of rest under its influence, and this would apply directly to the deduction of the laws of a body constituted as a pendulum under its influence ; and hence the laws of the pendulum, as actually moving under the influence of terrestrial gravitation, might be said to be deduced *from pure theory* and the abstract idea of a constant force.

But the real application of such reasoning essentially involves the actual existence in nature of such a force as that of gravity, which can only be derived from observation.

If the deviation of Foucault's pendulum had been originally a mere matter of *observation*, it would have been long before experiment would have arrived at the solution. Many would have been the hypotheses of peculiar magnetic, electric, or other causes, for the observed deviation. _{Foucault's pendulum experiment.}

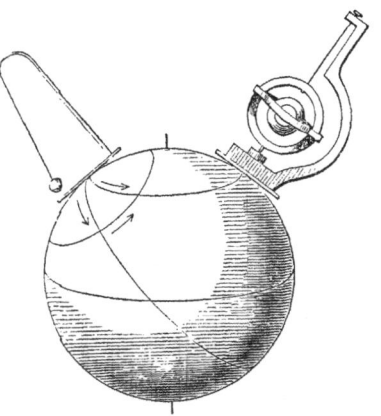

FOUCAULT'S PENDULUM AND GYROSCOPE.

It was only from a just mathematical conception of the *resolution of the rotatory motion* of any point of the earth's surface into two,— one *round that point*, the other at right angles to the former, and which would not affect the plane of the pendulum's vibration, while the former would,— that M. Foucault foresaw the result.   But this was *not à priori* reasoning disclosing a physical fact; it was simply reasoning deductively from a known fact to a consequence; when the reasoning being logical, that consequence could not but be true and be confirmed by observation.

Paradoxes of the gyroscope. Yet more astonishingly paradoxical are the effects exhibited by means of the gyroscope, which seem to subvert all the acknowledged principles of equilibrium.   To mention one only : a wheel loaded round its circumference, in rapid rotation at one end of a horizontal axis, having the other end merely resting on a pivot, is *supported on that pivot alone against gravity*, the whole at the same time revolving round the pivot.

Scarcely less remarkable is the application of this instrument by M. Foucault to another manifestation of the earth's rotation : — the wheel retaining its original plane of rotation, which therefore *apparently* deviates with the rotation of the earth.

It is probable that any person, even of considerable mechanical and experimental knowledge, seeing the action of the gyroscope for the first time, would be much puzzled to account for it, as, in fact, several persons have been; and if he set about investigating it experimentally and *inductively*, might be long before he traced any law or connected it with any principle, so as to reconcile it with the established doctrine of equilibrium.

If, however, he set out with a mathematical knowledge of the principle of the " composition of rotatory motion," and proceeded *deductively*, the explanation is easy, and its relation to a number of other important cases readily manifest. Yet the application of this mathematical theory requires the idea of a *material* body in rotation.

The ancients, notwithstanding all their refined geometry and spirit of abstract speculation, were unable to advance to the solution of the case of oblique equilibrium, or the inclined plane; and this is clearly a case where, if anywhere, *à priori* principles would have availed. But it was not until Stevin reasoned, *not* upon any abstruse axioms, but on simple mechanical considerations, that the demonstration was discovered. The solution was effected

Inclined plane.

D

by reasoning *deductively;* but it was deduction from principles obtained on primary physical or experimental properties of matter.

Discovery of magneto-electricity.   A highly instructive instance of the application of an abstract principle to physical discovery may be found in the way in which Faraday reasoned to the discovery of magneto-electricity, which I cannot describe better or more briefly than in the words of Mr. Grove * : —

" The discovery of Œrsted, by which electricity was made a source of magnetism, soon led philosophers to seek a converse effect; that is, to educe electricity from a permanent magnet. Had these experimentalists succeeded in their expectations of making a stationary magnet a means of electric currents, they would have realised the ancient dreams of perpetual motion — they would have converted statics into dynamics — they would have produced power without expenditure; in other words, they would have become creators. They failed, and Faraday saw their error; he proved that to obtain electricity from magnetism, it was necessary to super-

---

* Lecture on Progress of Science; London Institution, 1842, p. 20.

add to this latter, motion; that magnets, while in motion, induced electricity in contiguous conductors; and that the direction of such electric currents was tangential to the polar direction of the magnet— that as dynamic electricity may be made the source of magnetism and motion, so magnetism conjoined with motion may be made the source of electricity. Hence originates the science of magneto-electricity, the true converse of electro-magnetism."

The application of mathematical reasoning to physical inquiries may sometimes, at every step, exhibit something corresponding to an actual step in the mechanical process, and thus capable of a physical interpretation: such is often the case in the older geometrical investigations. But in the prevalent applications of the modern analysis there is no correspondence of this kind; the original conditions being once put into an equation, we resign ourselves to mere symbolical operations, which have individually no reference to any physical ideas, till we find ourselves landed as it were on the platform of a conclusion which marvellously harmonises with experimental results.

Yet these and the like instances are not at all

cases of an *à priori* discovery of physical truth; they
are instances of a train of logical reasoning proceeding from some first principle derived from remote
physical abstractions till it arrives at a conclusion
which coincides with some other observed law having
no other perceptible connexion with the first principle; or leads the philosopher to expect such a
result; which, on trial, is found to be the fact.

Conclusion.
No really
à priori
proof of
physical
truth.

Thus a simple analysis of the actual train of argument tends to dispel the mystification and confusion which have sometimes arisen on the subject
of abstract reasoning applied to physical subjects.
Pure reason out of its own resources may, indeed, create theories apart from all observation of
nature; but, to make them applicable to anything in
nature, such creations of the mind must necessarily
and universally involve *some* small assumption of
material properties or mechanical conditions; which
can only be in some form or another ultimately
derived from observation: what is borrowed may be
very little, but it must be *something*; and it is a point
of interesting research to the philosopher to endeavour to ascend to the fewest and simplest possible
of such first principles.

A confusion of ideas is sometimes introduced by *Necessary truths only necessary consequences.* the use of the term "necessary" dependence or " mechanical necessity ; " as if it were a *blind or fated necessity ;* but what we mean is, *a necessity of reason or logical sequence.* It is evinced by the dependence of a series of ideas deductively followed out ; which are also found to accord in their result with natural facts and more comprehensive laws.

The subject here discussed, is beautifully illus- *Speculative ideas of Œrsted.* trated by the philosophical views broached in a posthumous work*, which has so fitly and honourably crowned the labours of the great Œrsted, and added a new claim to our admiration of his genius. In those essays he maintains repeatedly the proposition that "the laws of nature are the same as the thoughts within us;" "the laws of motion are such as are required by our understanding;"† "the law of the inverse square of the distance is a conception of reason;" and several like instances: all which I should fully admit, subject to the qualification above suggested and understood in the sense

---

* "The Soul in Nature," translated by the Misses Horner.   London, 1852.
† See especially pp. 10. 36. 93.

which it implies—that the connexion and dependence of the facts in nature accords with the connexion and dependence in our reason, *provided* we
set out from some more or less simple principle
originally *derived from observation*, whence we advance by abstract reasoning to a conclusion, which,
however remote from the physical point whence we
started, is found to accord with natural facts, and to
be a general law of nature.   In this sense I have
before considered some of the cases just mentioned;
and others adduced by Œrsted are more obviously
of the same kind; such as the lesser planetary and
lunar perturbations, too small for observation alone
to detect, yet indicated by theory; the identity of
lightning and electricity; the discovery of the metallic bases of the earth; all anticipated by theory; to
which might be added Œrsted's own grand discovery of ELECTRO-MAGNETISM, and that of the
planet Neptune in our own day.   But these cases
are, after all, not precisely in point to the original
question, since here the starting-point was obviously
previous *inductive* knowledge.

Accordance   These distinctions are important to the funda-
of reason
and nature. mental analysis of our reasonings on which we ad-

vance legitimately to those broader ulterior reflexions on which Œrsted enlarges, and which are the same to which the whole of the present inquiry points.

Œrsted has well remarked that it is a common error to imagine matter something constant and invariable. But the permanence and invariability of nature are not found in its individual parts, which are all undergoing perpetual changes. The invariable, he argues, is found only in the abstract nature of things: "nothing is invariable in nature but *laws* which may be called the thoughts of nature."*

Natural combinations (Œrsted observes) which appear accidental are not really so. "All effects obey natural laws; these laws stand in the same necessary connexion as one axiom in reason to another: that this combination is precisely a combination of reason we learn from this, that by reason we are enabled to deduce one law of nature from the other, and by the known laws to discover new and unknown ones. Innumerable as are the effects determined by natural laws in every object in nature, however insignificant it may be, I deeply feel an

---

* The Soul in Nature, p. 23.

unfathomable reason within them, of which I can only comprehend by fragments an incalculably small part. In short, nature is to me the revelation of an endless living and acting reason."

" If the laws of reason did not exist in nature, we should vainly attempt to force them upon her: if the laws of nature did not exist in our reason, we should not be able to comprehend them." *

And on the whole, " we find an agreement between our reason and works which our reason did not produce." . . . " All existence is a dominion of reason." " The laws of nature are laws of reason," and " altogether form an endless unity of reason," . . . " one and the same throughout the universe."†

---

* The Soul in Nature, p. 18.   † Ib. 12. 16. 87. 92. 377.

§ II.   THE UNITY OF SCIENCES.

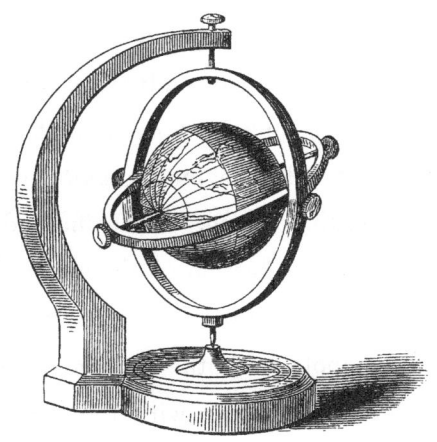

ALL branches of inductive science continually tend more and more towards a grand *unity of principle*. We perceive this to a partial and limited extent in every lesser advance of discovery: in proportion as new facts accumulate and become embarrassing from their multiplicity, sooner or later some happy advance in generalisation is always found to occur by which they are simplified and reduced to some

Tendency of sciences towards common principles.

single principle, connecting them at the same time with other classes of phenomena.

Gradual progress of generalisation and union of sciences.

In the science of the ancients (exact as it was in some limited departments, each within itself) all branches were isolated and disconnected: and all physical principles and causes were supposed of separate and even conflicting kinds.

First advances.

All the first great modern advances were directed towards *combining* and *uniting* branches hitherto distinct, and tended to evince a *unity* of idea and principle pervading them. The first discoveries pointed to the identification of the celestial motions with terrestrial; of astronomy with mechanics; of the fall of an apple with the motion of the moon; of the horror of a vacuum with the laws of equili-

Later physical discoveries.

brium: as later discoveries have identified magnetic and electric currents, and connected sound, heat, and light with the mechanism of waves; and, again, the resulting effects of heat with dynamical force.

Faraday's generalisations.

Of the tendency and progress of discovery towards a coalition and combination of different trains of research, perhaps we can nowhere find more striking instances than in the multitudinous researches — and every *research* terminates in a *discovery* — of FARADAY. The peculiar character of high generalisation which results out of an apparently immensely complicated mass of small details, is perhaps one of the most striking features of this wonderful series of investigations. It is impossible here to do more than select one or two instances.

Few generalisations of a more striking character have ever been announced than that of the *magnetic properties of all matter*, evinced in the classification of all substances under two species, magnetic and diamagnetic, and these characterised respectively by the properties of attraction and repulsion. *Magnetism and diamagnetism.*

But in this *union of relation* between magnetic and all other matter, there was to be disclosed a yet more striking instance *of bringing together remotely separated* kinds of physical action under a common law, in the action of *magnetism on light.* *Action of magnetism on light.*

What could be a more singular and striking *identification* of properties in cases *apparently* the most *remote* from each other than the production of *rotatory polarisation* in light passing through *quartz* and some other substances, and in passing through ordinary *transparent media when placed in the line of intense magnetic force ?*

Or to go back to an earlier discovery:

Definite electrolysis.

Grand indeed was the conception of the principle of the relations of *chemical* to *electric* action partially illustrated in theories of Fabroni and Wollaston, but first announced in all its generality by Davy: thus bringing into *close relation and unity* two such great modifications of physical power. Equally important, though apparently remote from either of the last was the principle of *definite proportions in atomic combinations* disclosed by Dalton.

These two comprehensive generalisations, each equally wonderful in itself, yet seemingly unconnected, it was reserved for the penetrating genius of Faraday to place in intimate connexion and to *unite in a still higher bond of generality.* No single discovery perhaps could be cited of higher intrinsic value than the disclosure of the great principle of DEFINITE ELECTROLYSIS: but the high philosophical character of this discovery is enhanced the more specially in that it *combines in a principle of unity* the mathematical *law of definite proportions in chemical combinations* with the preservation of the

same identical *numerical relations* in *electrolytic action :*
and thus uniting both in intimate relation with the
fundamental conception of *atomic composition.*

As we look to the larger divisions of the sciences,
and the successive wider generalisations which they
imply, the same tendency to unity is continually
though more slowly manifested.   And thus, even
where it does not yet appear, we cannot doubt that
this is the legitimate and ultimate direction and
tendency, however remote, of all scientific progress.
But in treating of the sciences systematically, it is
necessary to adopt some principles of classification
and arrangement.   Here some division is rendered
necessary for this particular object ; but it ought to
be carefully borne in mind that it should in no way
really interfere with the increasing conviction of a
real unity of principle pervading all branches.

Systematic division of sciences.

It is a reversal of the order of inductive advance
to endeavour to isolate each department of science,
and to place it on a separate base, by a theory
which would assign to each branch certain real
differences of principle and peculiar fundamental
ideas essentially characterising it.   If such a distinction were made out, it could be but a tem-

Some views tending to isolate sciences.

porary and provisional ground of classification, in time to be superseded by a reduction to a higher common principle.

Mechanical force.

It is no doubt true, that the highest, the most perfect, and satisfactory assignment of physical causes is effected when the phenomena can be analysed into *mechanical* laws. But the reason of this lies in no mysterious connexion of mechanics, as such, with the idea of causation, but merely in this, that the conditions of purely mechanical reasoning are so perfectly elementary in their nature, and so entirely free from all admixture of ambiguous or doubtful conditions, that we can directly investigate them with a simplicity differing in nothing from that of primary geometry, and thus attain the most perfectly satisfactory explanation, when everything is reduced to simple consequences of mechanical equilibrium or the composition of forces.

Progress of all sciences towards the idea of mechanical force and motion.

In other branches it is clear that just in proportion as we can succeed in reducing the phenomena from obscure and apparently mysterious modes of action to these simple and intelligible cases of force and motion, in the same proportion we bring those branches into the domain of exact science, and break

down the line of demarcation which hitherto seemed
to separate them.

The sciences of statics and dynamics, of equilibrium
and motion, have been represented by some writers
as based on inherently distinct principles : but it is
at once a more satisfactory, and as I believe a more
true, view which connects them by the consideration
that the simplest cases of *equilibrium* or *rest* cannot
fully be demonstrated without an explicit or tacit
reference to the idea of *motion* * : which thus far
helps the more general consideration of the ultimate
unity of all sciences.

The explanation of the precession of equinoxes
(the same in substance as that of Newton, more
circuitously followed out) by the direct application
of the composition of rotatory motion announced by
Frisi, and imitated by the rotatory apparatus of
Atkinson and Bonenberger, exhibits a peculiarly
striking exemplification of unity of principle in
passing from such phenomena, vast in their relations
both to space and time, to the identical cases pre-

Common grounds of statics and dynamics.

* See my "Essay on Necessary and Contingent Truth," Ashmolean
Memoirs : Oxford, 1847.

sented in the deviations of rotatory projectiles *,
the cases of spinning tops in stable and in unstable
equilibrium, and the various paradoxical effects produced by the gyroscope: all, however diverse, direct
consequences of one simple law.†

Polarity re-
duced to
resolution
of motion.

The idea of " polarity," to which such mysterious
importance has been attached, has been sometimes
imagined to involve some essential peculiarity supplying an appropriate characteristic conception to
mark a distinct class of physical phenomena. But
this once marvellous notion, in the instance of *light*,
has *been reduced to a simple case of resolution of motion ;*
and there can be as little doubt that the progress of
inductive generalisation, and the application of mathematical principles, will, sooner or later, reduce other
instances, at present provisionally designated by the
same name, to equally simple modes of action.

Dynamical
theory of
waves.

And with respect to the phenomena of optics
generally, how completely remote do they appear
from all notions of mechanical force? yet, by the

---

* See a Memoir by Prof. Magnus, translated in Taylor's Foreign
Mem., N.S. pt. iii. p. 210.

† The Vignette at the head of this Section represents the apparatus
as constructed on a large scale for lecture illustration at the Royal
Polytechnic Institution.

mathematical labours of Fresnel, Cauchy, and others, these seemingly remote appearances have been connected with a recondite theory of pure dynamics; which, followed out through a complicated train of deduction, ends in reducing nearly all these phenomena to the results of certain minute *motions*, subsisting and excited among a system of imaginary molecules acted on by attractive and repulsive forces, and subject to external agitation.

So, again, when electric and magnetic action were reduced to systems of currents by the researches of numerous and distinguished co-operators, in following out the great principle disclosed by Œrsted, there was a direct approach to ideas of *motions in definite directions*, which supply the abstract indications of force; and though the subject has even yet been but imperfectly followed out, we perceive the direction it is taking, and must eventually take, towards satisfactory explanation, in a reduction to simple dynamical principles. *Electromagnetic currents.*

One of the most remarkable approaches (as yet quite in obscurity) which has been made towards a connexion in principle between two branches of science apparently remote, is that of a peculiar *Supposed case of interference in galvanic action.*

E

action of a galvanic current exhibiting all the
marks of a case of INTERFERENCE, in the experiments of Mr. Grove.*   If this should be followed
out by a more close analysis, so as to show a real
action of the kind, the analogy of galvanic action
with *a system of vibrations* of a fluid analogous to the
luminiferous æther as its cause, would open the way
to a generalisation of the highest and most valuable
kind.   And further, it may not be altogether inconceivable that two sets of such vibrations, which, by
superposition, give rise to elliptic vibrations, may be
connected with the formation of currents running
round the wire, by which so many of the phenomena
are represented.

Molecular
forces.

Again : to insist on an essential scientific distinction between *molecular* forces and those acting on
matter in larger *masses*, as the characteristic basis of
a peculiar science, tends to isolate this branch from
ordinary dynamics, to which we should rather seek
to assimilate it.

In the same way the broader distinction between

---

* Phil. Trans. 1852. Part I.   Some highly interesting facts, *apparently*
corroborative of this idea, though possibly a different case, are described
by Dr. Robinson of Armagh. Proceedings of Royal Irish Academy, Jan.
14. 1856.

mechanical and *chemical* action tends equally to Chemical forces.
break up the idea of that essential and fundamental
unity which the philosopher is persuaded must really
subsist between these invisible actions of atoms on
atoms, and those more obvious, only because on a
larger scale, of worlds on worlds.

The distinction of *molecular* forces, there can
be no doubt, marks merely a *present* line of demarcation from ordinary mechanical forces, which
will at some future time be effectually broken down,
and the two classes reduced to one higher genus.
Chemical action, again, we may be assured, differs
from *mechanical* only in our existing state of ignorance ; but they will doubtless at some period be
assimilated by the discovery of a common principle of
equilibrium and its disturbance. Even in the present
state of our knowledge, molecular forces have been
shown with great probability to be reducible to a
common theoretical expression with that of gravitation in the speculations of Boscovich and Mossotti.

Again, the mode of aggregation of many of the Cosmical forces.
stellar clusters, as described by recent observation, is
regarded by some very eminent philosophers as
evincing the action of forces of a peculiar kind

different from those of gravitation.   Should this
prove to be the case, it would in no way derogate
from the universality of *some law* of aggregation of
matter, that a different species of law may prevail in
those vast distant portions of the universe, which,
when it shall have been investigated, may prove a
more comprehensive kind of force, of which gravitation is but one form or modification.

But if any such apparently outstanding exceptional
case were fully made out rightly to claim the title of
involving an entirely new principle, still the inductive
method would only mark out that principle as a
legitimate subject of future analysis ; and we might
be assured that in the successful course of such analysis at some future period, either this new principle
must fall under some already recognised principles,
or those recognised principles must fall under it.

There may no doubt be a practical convenience
in retaining some distinctions of this kind to *preserve
arrangement* in our subjects ; but to attempt to fix
them as essential foundations of real philosophical
distinctions, seems to be reversing the proper order
of inductive inquiry.   *Provisional* and *temporary*
distinctions for classification, indeed, we may with

convenience and advantage often make between different branches of science in regard to the modes of reasoning and nature of the leading ideas appropriate to them; but it is essential to remember that these distinctions are *only* provisional.

But in contemplating the unity of sciences, an exception has been alleged in reference to GEOLOGY. The entire relation in which it stands to other branches of *inductive* science, and even its *inductive character* altogether, has been sometimes disparaged. Comte has denied it any place whatever in the scheme of " positive philosophy," * and possibly some hypotheses which have continued to be occasionally indulged in, in connexion with that science, might not unnaturally have influenced him in entertaining a prejudice against it.

Alleged exception as to geology.

Yet this science, when rightly pursued, is eminently *inductive*. From its very nature it combines the resources of a variety of other sciences; dynamical, hydrostatical, chemical, and especially physiological, and being thus entirely dependent on these other branches of inductive philosophy, itself acquires a perfectly strict *inductive* character.

Not real: geology an inductive science.

* See Appendix, No. XI.

E 3

When, at the present day, it exhibits to us, preserved in their stony sepulchres, the successive varieties of organised structures, as they lived and moved
in the same world, subject to the same immutable
laws, mechanical, optical, and physical, uninterruptedly in operation through all the incalculably vast
periods of past time, it is an entire departure from
all just appreciation of the unity of science and of
nature to imagine that any essentially different laws
of vitality then prevailed, or that the changes in
organised life thus brought to light were governed
by any totally different series of causes from those
now in operation of a peculiar and mysterious kind.

Yet some seem to have supposed that the reasoning of geology ought to rest on something distinct
from that of the experimental sciences, — inasmuch
as it refers to events which have so long since passed
away, and which we cannot recall for examination,
while the very terms "*palæozoic*" and "*palætiology*,"
might seem to insinuate that we are concerned with
an order of causes belonging to the past, different
from those now in action, — a distinction just as
unphilosophical as that of the peripatetics, who drew
a distinction between "natural and violent" motion,

No really different causes referred to.

The evidence of geology not different in nature from that of other sciences.

and ascribed the terrestrial motions and the celestial to distinct causes.

Induction has no reference to distinction of *past* or *present;* if phenomena have been locked up for ages, yet, when once thrown open to us, they become objects of the same kind of investigation as those occurring at the present day. The investigation and restoration of the remains of a Saurian imbedded millions of ages ago, is an operation of precisely the same kind as the *post-mortem* examination of the subject of yesterday.

The inductive philosopher is convinced that the universal subordination of causes must hold good equally in *time* as in *space;* that as there is no *region*, however distant, in which physical laws do not apply, or in which, if as yet unknown, we are not fully warranted in feeling an assurance that they must apply; so in *time* there is no *period*, however remote, at which we can legitimately imagine the chain of physical causation to be broken, and to give place to disconnected influences of a wholly different kind.

Uniformity of nature in time and space.

More recently, the investigations of Mr. Hopkins have tended to connect geology even with dynamics

Geology approximating to an

E 4

exact
science.

and mathematical laws, and thus to establish its relation, not merely to the *inductive*, but even to the *exact* sciences: not that that name implies any real difference in nature, but merely marks the *degree* of perfection to which any branch of science has attained. If, then, from the examination of phenomena *actually existing*, and going on around us, we turn to the past, the rules and principles of inductive investigation will apply with equal force and propriety to phenomena which teach us the successive and gradual changes which the crust of the globe has undergone, and lead us to trace them as far back as we can towards its origin.

Influence of
time ad-
mitted.

The great principle which forms the basis of all inductive geology — the analogy of existing causes in explaining past changes must, however, be distinctly understood, and, in fact, is so interpreted by its best advocates, not merely as restricted literally to those identical natural operations which *we see going on*, AND COMPLETED, daily before our eyes within the limited moment of time to which our observation extends.

It would not fully vindicate its own power, if it did not include in the general analogy the influence of some elements *incapable, from their nature of direct verification from our own experience,* such as are due to the INFLUENCE OF TIME, especially of *unlimited periods of time;* and in illustration of this idea we are reminded of some changes even in more limited periods, which, though in their nature and results simply chemical, are yet *such as cannot be,* or at least *have not been, produced in our laboratories.* We may take as instances the formation of *coal* and of *diamond;* while on a grander scale we are under the necessity of acknowledging the long series of changes which must have accompanied the gradual cooling of the earth, an unavoidable inference from the fact of existing central heat.

Illustration from formation of coal and diamond.

Real inductive principles thus tend to reduce to order those phenomena which have appeared to some to present so much more strongly marked vicissitudes only because we are apt to crowd the events together in the long perspective, and measure them too much according to our confined ideas of duration.

Misconception of past duration.

In speculations on changes where, it is alleged, all applications of known causes fail, it has been the favourite resource with some to appeal to mys-

Theory of convulsions.

terious revolutions and occult operations of a kind
ill-explained, and even supposed to be inscrutable
to our faculties, but thus the better calculated to
dazzle the many with their imposing pretensions.

Of an unin-
ductive
character.

But in the spirit of true induction we have no
right to imagine that any of the events or changes
of past epochs, however apparently inexplicable,
can be rationally set down as events of a *different
kind and order* from those now going on, or as interruptions of the settled order of natural causes.

Uniformi-
tarianism
and cata-
strophism.

Difference of opinion indeed may subsist as to
the *greater or less frequency* or *intensity* of volcanic
action, of fractures and dislocations, of variations
in climate, of changes of condition due to the
cooling of the terrestrial nucleus, or the like, in
past epochs. But these, while they are on all
hands allowed to be fair and legitimate topics of
philosophical debate and inductive inquiry, would
be most unduly exaggerated if supposed to mark
any such real or fundamental difference in principle
as to constitute two really distinct geological schools.
They are questions merely of *degree*, not of *kind* or
of *principle*.

Yet, in the language often used, the "uniformitarian" view would seem to be represented as an hypothesis to be fairly weighed against another antagonistic " catastrophic " theory. If the terms are to be understood with any such difference of sense as that thus implied, I conceive it appears that the two theories respectively occupy totally different grounds.

The " uniformity " principle would mean simply the proper extension of inductive analogy and the law of continuity, even if not yet sufficiently substantiated in detail in each particular instance; while the " catastrophic " hypothesis seems of an essentially uninductive nature, and appeals to ideas remote from true analogies, confessedly resorted to on the very plea of the failure of explanation by natural causes.

But, in such cases, the evidence of a violation of the uniformity of nature is purely *negative:* with all analogy against the reality of the exceptions, they *can be such only to our present ignorance:* the apparent anomaly is but a part of a more comprehensive law, ill understood; — a modification of its continuous action in reality equally regular, though not as yet fully made out or reduced to

law.   Geology thus kept pure from the introduction
of fanciful and unphilosophical hypotheses eminently
conforms to the type of unity which binds together
the whole range of inductive science.

Revolutions
in science
only pro-
gressive.

The unity of sciences is not impaired, but insured
and promoted, by those *mutations* which any of its
branches may seem to have undergone.   All real
science is in a state of perpetual change.   These
changes have now and then been fundamental and
revolutionary, and similar fluctuations are perpetually
going on in lesser details.   But this in no way makes
science itself unstable or fluctuating.   The change
is always of one character, and that no other than
the very nature of the inductive philosophy requires :
a change from anomaly to regularity, from hetero-
geneity to analogy, from confusion to order, from
interruption to continuity, from artificial dogmatism
to the simplicity of nature.

Discoveries
superseded
only by
greater im-
provements.

Every branch of science approaches perfection and
stability as it more fully approaches to and realises
the grand principle of *unity*.   It is the test of the
real advance of discovery to exhibit a progressively
increasing conformity to these great principles : an
advance which will not require a retreat, — the

erection of a structure which will not require re-modelling.

Every philosophic research or conclusion, at present of the highest importance, must expect to be reduced to a subordinate place: every method now most justly esteemed must look to be superseded by greater improvements: but nothing will deprive such really great discoveries of their place in the page of history—their lustre will but be increased by the brilliancy of newer results, to which they were the necessary preliminaries.

Such mutations are sometimes made a topic of reproach, but only by those who are hostile to science from entire ignorance of its principles; they may learn to observe that these changes are *all in one direction*: they are all steps in advance towards a higher and more enduring system—all future progress must be in the same direction; we shall never see a recession from the more natural towards the more mysterious; from the recognition of regulated causes, law and order, in a retrograde course towards arbitrary or fortuitous influences.

In the study of nature all things are at first presented to us in an obscure and mysterious form,—their

Advance from mysticism to reason.

shape and outline is but dimly seen,—we are lost
amid a mass of heterogeneous elements of crude
materials, unconnected facts, disordered imaginations :
the first attempt is to refer them to imaginary and
mysterious relations, occult virtues, efficient causes,
sympathies, antipathies, affinities, polarities.   But all
this mystery and confusion, it is the very business of
sound philosophy to analyse—to clear up by luminous
distinctions, and reduce to intelligible conceptions ;
and though some sound philosophers may continue
provisionally, or for convenience, to use some of the
same *terms*, yet they carefully distinguish them as
nothing more than terms of convention, however
inferior apprehensions may be misled to mistake them
for realities.   So long as mystery continues to haunt
us we have not really entered on the domain of philosophy — where science begins there mystery ends.

To recur then to mystery as the end of philosophy,
is to invert the order of things.   But there is nothing
at variance with this rule in tracing the indications of
mind, which necessarily result from the manifestation
of design and harmony in those universal laws
which are the very clearing up of physical mystery.

Applied to several sciences.   Most sciences had their origin in the clouds of

mysticism, and thus occasionally long retain some
tincture of it.   Astronomy arose out of astrology,
chemistry out of alchemy, and geology out of a
theological cosmogony.   Geology, indeed, being the   Advance of
youngest of the inductive sciences, has naturally   geology.
in the course of its rapid growth, within a brief
period, exhibited more of those changes from mysticism towards rationalism than any other branch.   It
is but a short time since the whole science consisted
of little better than a few detached general facts,
connected by arbitrary hypotheses, and conformed
to the language of dogmatic belief.

With an increasing recognition of true inductive
principles, we have witnessed progressive improvements in the philosophic character of the theory and
candid retractations of opinions once warmly upheld,
chiefly on grounds alien from those of science.   Yet
these concessions perhaps were made more from the
disclosure of a few contradictory facts in particular
instances, than from any perception of broader philosophic principles as those which in the first instance
ought to have formed the basis of the whole science ;
and, perhaps, such principles are hardly yet universally recognised in their full force and extent.

Influence
of dogmas.

Those who continue really to indulge in the visions which misled geology in its infancy, the dreams of universal cataclysms, and sudden creations, of a kind wholly remote from physical analogies, and to which it would be wrong to seek to apply physical explanations, so far place their speculations out of the pale of the inductive philosophy.

But the influence of such artificial theories we may be assured will in time entirely disappear, and all true cultivators of science will come to regard such distinction of schools in no other sense than as we now speak of Ptolemaists and Copernicans, Cartesians and Newtonians: these anticipations, however, are far from being yet generally realised. Many who smile at the fancies of a Whiston or a Buffon are scarcely less under the dominion of ideas of very kindred origin. Those who disown dogmatic authority to teach the mode of formation of the earth's crust are yet often not exempt from prepossessions equally narrow in speculating on the probable order of creation, the succession of species, or the relations of our globe to other planetary and stellar worlds.

Tendency

But to minds duly impressed with the great principles

of analogy, law and order, all anomalous imaginations to principle
of unity.
derived from sources extraneous to science will disappear.   The increasing tendency of all research
towards harmony, simplicity, and unity of character,
will be recognised as a pledge of its ultimate rea-
lisation : and even conjectural hypotheses, confessedly a mere indulgence in philosophical romance,
provided it *be* strictly philosophical, will be hailed
with satisfaction as helping out the general conception and keeping alive the spirit of analogical
inquiry.

But a yet more serious question, of the same kind Second alleged ex-
as that referring to geology, has been raised with ception in
the science
respect to the *sciences of organisation and life :* which of organisation and
are sometimes supposed to involve altogether a *new* life.
class and order of ideas of so peculiar a kind that
they must stand out as entire exceptional cases to
the general *unity of the sciences.*

Now it will on all hands be allowed that these Peculiar
difficulties
subjects are as yet but imperfectly understood, and of the subject.
a large range of inquiry connected with them still
involved in obscurity.   And if from external phenomena we seek to advance to their causes and principles, it is of course most fully admitted that of the

F

ultimate causes of organisation and life we cannot
at present attain to any satisfactory explanation, or
even form any definite conception.

But hence we find many in treating the subject
commonly set it down as in its own nature *something
essentially mysterious and inscrutable*: as referring to
an order of causes altogether distinct, wholly disconnected with those of any branch of physical investigation; as involving functions and operations
wholly *sui generis*: and not only that we *cannot*
explain them on any merely physical principles, but
that we *ought not* to attempt to do so: that they
are of an order wholly transcending such inquiries;
beyond the power of our faculties to apprehend;
and ought to be kept apart, as being indications
of a special and mysterious principle which it would
be presumptuous and immoral to attempt to inquire
into.

Not really
mysterious
or inscru-
table.

Everything doubtless is mysterious till it is made
known, but the inductive inquirer will never allow the apparent obscurity of a subject to oppose
any barrier to the endeavour to make it clear.
Nothing can be more *mysterious* than gravitation;
but that does not hinder the philosopher from in-

vestigating its laws, or thence, as far as he càn, penetrating towards its principle. Electricity and magnetism, thunder and lightning, were perfect mysteries a century ago. Instead of allowing any such prepossessions to paralyse his researches, the inductive philosopher would simply seek in regular order, first to determine the *external* conditions and *laws* of life, themselves as yet far from being well understood. Until these are known, he might reject as premature, or at least regard as wholly conjectural, all attempts to speculate on their higher laws or physical causes: yet not less confidently would he be assured that these more interior causes will one day come to be known; just as surely as the proximate laws will be accurately traced and reduced to that determinate order which undoubtedly in reality pervades them, but of which we have at present only the most imperfect glimpses, yet which, imperfect as they are, are the true openings to the ultimate inductive knowledge of causes and principles.

There have not been wanting, indeed, attempts at theorising on the subject: various hypotheses have been started as to the nature of the " vital principle,"

Proposed hypotheses of the vital principle often fallacious.

and the question discussed whether life is the result of organisation, or organisation of life. Some have referred to more particular modes of action, such as electric currents flowing through the nervous system, or the like; and have represented animated beings as in fact nothing more than walking galvanic batteries; all these, and many similar theories, *may* be utterly fallacious and erroneous; and the opponents may triumph and revel in the real or supposed refutation of them. But all this in no way affects the conviction of the existence of *some physical principle*, the cause of the vital functions, as yet, indeed, unknown, but which nevertheless will, at some time, become as well determined as the principle of respiration or the circulation of the blood are at present.

Again, though chemical analysis has reduced organised products to determinate elements, yet it is made a matter of no small boast by some, that no chemistry can reproduce an organic substance, or invest that organised substance with life: and eager and loud was the triumph of those who conceived they had refuted the alleged results of Messrs. Crosse and Weekes, and bitter the abuse and ridicule heaped upon them for believing that they had evolved insect-life by galvanism.

All such experiments *may* indeed be fallacious and premature; and we *may* be as far as possible from at present penetrating the secret of vitality, or the precise mode of its connexion with the bodily structure and the chemical changes elaborated by the various organs. But the truly inductive inquirer can never doubt that there really exists as complete and continuous a relation and connexion of *some kind* between the manifestations of life and the simplest mechanical or chemical laws evinced in the varied actions of the body in which it resides, as there is between the action of any machine and the laws of motion and equilibrium, — the weaving of cloth by a power-loom and the principle of latent heat: and that this connexion and dependence is but one component portion of the vast chain of physical causation whose essential strength lies in its universal continuity, which extends, without interruption, through the entire world of order, and in which a real disruption of one link would be the destruction of the whole.

The principles of inductive science apply to all

But some physical cause of life.

All nature
subject to
law and
order.

physical truth and the investigation of all *physical*
causes.   The laws of order, uniformity, and continuity belong to all parts of the *material* world:
and in this order and continuity *animal life* is included.   From the lowest mechanical or chemical
influences on inorganic matter, there is an unbroken
series to the first manifestation of organic changes;
and from these again—from the lowest vegetable
or zoophyte up to the highest mammalia — there is
entirely one continuous progression, its connexion
from one term to another being carried on through
absolutely insensible degrees and shades of difference.

Humboldt observes,— "All myths about imponderable matters and special vital forces inherent in
organised beings, only render views of nature per-
plexed and indistinct."*   It is the unbroken preservation of this continuity which assures us that the
nature of the vital principle must be sought for by
no occult or mysterious process, but only by the
patient application of the same inductive processes
by which other physical principles have been and

---

* Cosmos, 69. transl. 1845.

always continue to be gradually cleared up and elicited; and by the operation of which, we may be assured, this hidden spring of life will, at some time, be disclosed, and brought out to occupy its place in harmony with all the other great principles of the universal cosmos.

But there is another plea on which the phy- *Physiology supposed distinct, as founded on final causes.* siological sciences have been sometimes supposed to stand apart from other branches. It is alleged they are characterised by involving the peculiar and distinctive idea of *organisation,* that is, an idea essentially involving the conception of *design* or *intention,* and have hence been referred to a separate principle called *teleology.*

This, however, appears to me a distinction unfounded in itself, or rather founded on an *incidental* and not on an *essential* distinction, and referring rather to the narrower view of this class of investigations as followed by an older and less advanced school; whereas in their more modern extension, they imply a more enlarged principle, and one closely accordant with the extension of analogy and the unity of science.

It is of course obvious that throughout these

sciences, perpetual instances of such adaptation of structure to the ends and purposes of life are abundantly manifested; and it is no less evident that they force themselves on the mind with that peculiar, immediate, and irresistible kind of effect which is justly dwelt upon by most writers on the subject, and admitted by all inquirers in such multitudes of convincing examples. On these, however, it is not my object to enlarge here; the present question is as to the precise philosophic analysis of the case with reference to the classification of sciences.

So *rapid* is the mental operation by which the inference of design in these cases flashes upon us, and so *immediate* is the impression, that it may seem almost to *precede*, or at least to go hand in hand with observation, without waiting for formal deduction: so that we may not unnaturally deceive ourselves, and may sometimes mistake it for an intuitive notion, acquired antecedently to the actual examination of organised structures, and may even imagine (as some have even maintained on philosophical grounds) the idea of a purpose, an end and means, is an integral part of our very idea of an organised being. Yet

when we analyse our conceptions more strictly, it
must be apparent that our very notion of the existence of organised beings must be acquired in the first
instance from observation — including the observation
of ourselves: and this constitutes so constant and
universal a case of experience, that it may well seem
an idea whose origin we may set down as con-
temporary with our earliest exercise of consciousness
and thought.

It is, however, in strictness, not merely from
observation, but by a considerable exercise of inference and deduction, that we can legitimately
arrive at the notion that an animal "is intended to
live;" it is derived from the study of its organisation:
whence we are led to look to the subserviency of its
parts to the purposes of life and enjoyment.

Not essential, but incidental.

The idea which we form in general of an *organised
body,* no doubt practically involves that of parts mutually dependent and adapted to each other ; but this
is an *inference,* and the relation which it establishes is
one in no way *essentially* differing, in this respect,
from that existing among the component portions of
a moving machine, or even of a stationary arch;
though certainly differing in the *degree* of compli-

cation, as in the higher and more varied and multiplied design and object evinced in their structure.
There is no essential distinction *in kind* between our
conception of the one or the other. It is true we
soon come practically and habitually to include these
effects in the complex current idea of an organised
being, and are unconsciously and involuntarily led as
it were to connect these conditions with the idea of
plan and intention, and to assume the relation of these
as that of ends and means. But we are here concerned only with the *analysis* of our ideas.

More en-
larged views
necessary.

That a combination of arrangements, perhaps
even complicated ones, which answer a purpose
whose practical importance is obvious, and where the
relation of one to the other as end and means almost
forces itself on the mind the moment we contemplate
them, must produce a high conviction of design, is as
indisputable as it is invaluable in the high argument
of which it forms a part. But such instances arising
in the contemplation of *organised* structures do not
stand in any way *peculiarly* distinguished *in their
nature* from other cases of the like adaptation of
means to an end in the wider arrangements of *unorganised* matter.

It may readily be granted as the fact, that several remarkable physiological discoveries have been made in consequence of the habit of looking at final causes in animal structures. But what does this prove? Reduced to its proper place in the philosophical system, the case is simply this: most discoveries in physical science are originally prompted and suggested by some previous *conjecture*. Nothing can be more fruitful in furnishing such conjectures than the habitual recourse to instances of adaptation to an end in organisation already known, whence the enlightened physiologist often receives the most valuable hints and frames the most probable conjectures as to those which are as yet unknown. The value and force of such conjectures in general depends on the happy preservation of *analogy*; and that analogy is in these cases most likely to be traced in the connected series of means and ends.

Discoveries made by help of final causes; only as hints.

The object is not in this place to enter on the general argument of "final causes:" and in reference to the present subject I will only remark, that the wider extension of physiology by the introduction of the more enlarged and modern principle of "unity of composition," besides its proper claims

Unity of composition the true principle of philosophical physiology.

as the basis of all great and scientific conceptions of such subjects, is also remarkable in this respect, that it leads us more directly to recognise the proper place of physiology among the sciences as exhibiting it more clearly in its relations to that *unity* of prin-

Hence unity of physiology with other sciences. ciple which pervades them all. There is nothing *exclusive* or peculiar in the study of organised bodies ; it involves no essentially characteristic, idea distinct from other branches of physical investigation, but, like them, tends to the grand conclusion of a reference to common and high principles of unity and harmony of plan and design throughout nature.

Difficulty as to including man in the series of nature. But the most difficult, and at the same time the most important question in any theory of this kind, has been raised on the ground of its relation to *the nature of* MAN.

It will, however, hardly be denied that man, *considered in his animal nature alone,* is very little superior to brutes, and in some respects inferior. In the scale of *mere animal organisation,* the difference between the lowest human form and the highest monkey is not greater than between one class of monkey and another. Whatever difference of opinion may have arisen on this subject of a moral and metaphysical kind, yet it is on all hands allowed that man has *to a certain* extent a nature in common with brutes : and we may *avoid all cavil* if we simply assert that man, *in so far as* he partakes in a nature common to brutes, is along with them, in *that respect*, a part of the same scale and system of organised life. *In so far as* his animal nature, functions, and instincts are concerned, they are linked in the same chain of continuity with the order of other material existences.

Distinction of man's nature; animal and higher.

To *what extent* mind and volition, especially in their lower functions, in *man* are different from the corresponding manifestations in inferior animals, is doubtless a very important question of psychology. To draw the line may be difficult or impracticable. Without pretending to determine such a point, we may safely say that, *in so far as* they belong to the animal part of man's constitution, the question as to the nature of such manifestations of intelligence may be a question of *degree,* and may be philosophically treated as connected with other questions of man's physical development, as part of the great scale of natural existence, governed by natural laws as yet

very imperfectly known, but fairly subjects of inductive inquiry. The question of an intellectual
principle, in so far as it is of a *metaphysical* kind, can
in no way affects the continuity of man's *physical
nature* with the rest of the material order of things.

Man's
spiritual
nature of a
different
order of
ideas.

But the more important question refers to the
further assertion of a distinct *moral and spiritual
nature* or principle existing in man, and all the
higher relations consequent upon it, which place the
nature of man *in this respect* in a category altogether
different from that of inferior animals.

Now on this most important point I would only observe one thing in reference to our present subject :
the assertion in its very nature and essence refers
wholly to a DIFFERENT ORDER OF THINGS, apart
from, and transcending, any material ideas whatsoever : hence *it cannot be affected by any considerations
or conclusions belonging to the laws of matter or nature.*

Man con-
nected in
the natural
series.

In a word, man's nature and existence on earth is in
nothing of a *peculiar* kind, and in no way violates
the *essential unity and continuity* of natural causes : —
in regard to man's *animal* nature, because, *so far as
that extends,* it wholly belongs to the physical order
of things ; — in regard to man's *spiritual* nature,

because, so far as it is properly such, is avowedly independent of all material considerations, and is therefore relieved from all possibility of connexion, or collision, with any physical truths, or theories.

Man considered in his *animal* nature, and as a part of the physical order of things, is, beyond doubt, just as much subject to the universality of natural law and order, as any other portion of animated nature. But there are those who take great exceptions to assertions of this kind as considering them of a nature lowering the dignity of the human race, and degrading to man's superiority. I reply to such objections, by observing that man's superiority is in no way compromised, to whatever extent we carry such observations ; because his *real superiority* consists *not* in the *physical* but in the *moral and spiritual* part of his nature : and this is admitted by the objector to be of a kind altogether distinct and belonging to a higher order of things, not amenable to any physical considerations. Some, indeed, maintain that it is difficult to draw the line, or to say, what part of the complicated tissue of the human constitution can be properly said to be physical and what moral. And some even contend that the moral is more entirely absorbed and included in the physical than others will allow to be compatible with spiritual and religious considerations.

I do not enter on any discussion as to *how far* the physical and the metaphysical part are psychologically distinct. I would concede that they may be to any extent closely dependent upon each other or intimately combined. But to whatever extent we may advance, or recede, whether towards the more spiritualistic or towards the more materialistic view, still to refer to the consideration of thought, volition, mind, or spirit in a metaphysical, moral, or spiritual point of view, is professedly to enter upon a new world, out of the region of physical investigation and belonging to the province of a higher order of inquiry, with which that of physical causes has nothing in common. Thus to whatever extent the dominion of physical investigation may be pushed forward, still the realm of moral and religious truth remains uninvaded.

Similar distinction as to the past.

As, then, the foregoing consideration refers to the study of the *existing* relations of organised life and of man's nature, *so far as* it belongs to animal existence,

so the same principles equally apply to the investigation of its *past* history and origin, so far as we can
trace it. We need seek no more for peculiar or
occult causes in the one case than in the other.

If we admit that the earth, being still hot internally, must have cooled at its surface, and that this
cooling must, in its progress, have caused contortions,
dislocations, upheavals of strata; and again, that the
waters charged with matter must have deposited it;
and that the various crystallised bodies and metallic
veins must have been formed during certain stages
of these operations,—it is only by parity of reason
affirmed that the rudiments of all organic as well as
inorganic products and structures must have been
evolved in like manner, as they were alike included
and contained in the once fused, and therefore once
vaporised or nebulous, mass. In that mass all kinds
of physical agents, or the elements of them, thermotic,
electric, chemical, molecular, gravitational, luminiferous, and by consequence not less all organic and
vital forces, must have been included.

Rudiments of all phy-sical things in the pri-mæval mass.

Out of it in some way, by equally regular laws in
the one case as in the other, must have been evolved
all forms of inorganic and equally of organic existence,

G

—whether amorphous masses, crystals, cells, monads, plants, zoophytes, animals, or man,—the *animal* man; the *spiritual* man belonging to *another order* of things, a *spiritual creation.*

Conclusion.

From this brief discussion, which was rendered necessary in order to meet some apparent exceptions to the general view and assertion of *the unity of sciences,* we may now return to the main conclusion, equally valuable in regard to the view it tends to open of the study of the sciences and their relation to each other, as in its bearing on higher inferences which are the crowning pinnacle of scientific truth.

Unity of sciences represents unity of nature.

Sciences in different stages of advance.

All science then is emphatically one : in all its parts and branches, however apparently distinct, or supposed to involve peculiar modes of thought appropriate to each, we find, on close examination, that all such distinctions are but temporary and provisional, and that what appears peculiar is so only because the investigation in different parts of science is in different stages of progress. In one it has arrived at no more than a description and classification of phenomena, or even of the materials whose phenomena, we propose to study; in another we have been able to reduce all phenomena to laws of high generality, and those laws to simple principles of force and motion of the most elementary simplicity and the highest generality ; and between these extremes there exist all varieties of intermediate stages.

But all sciences approach perfection as they approach to a unity of first principles,—differently applied, indeed, according to the different nature of the material objects contemplated, but in all cases recurring to or tending towards certain high elementary conceptions which are the representatives of the unity of the great archetypal ideas according to which the whole system is arranged. Inductive conceptions, very partially and imperfectly realised and apprehended by human intellect, are the exponents in our minds of these great principles in nature.

All sciences approach towards showing the unity of nature.

The great inference of uniformity is corroborated not only by the successively more and more comprehensive laws of nature, which science exhibits, but by the very possibility of the existence of such a thing as systematic science : not only by the accumulative proofs existing in nature, but by the marvellous adaptation and harmonising disposition

of the human mind for appreciating and discovering them : not only by the occurrence of natural
events in invariable order, but also by the possibility of expressing them by laws conveyed in
exact terms, and of advancing deductively to the
prediction of other phenomena. Thus, even this
preliminary condition of all inductive inquiry affords
confirmation of the principle of unity of design,
connecting the physical with the intellectual world ;
and this in a still higher degree, as all sciences are
seen to tend towards unity.

Our con-
ceptions of
natural
order the
reflexion of
the reality
in the su-
preme
mind.
The actual laws and profound principles which
regulate the mechanism of the universe are the
originals, the conception and expression of them in
the mind of man, only the copies. The vast assemblage of physical causes, whether the great principles
of cosmical forces, or the minutest molecular affec-
tions, — as they exist in the heavenly spaces or
among terrestrial atoms, are the realities : the exposition and demonstration of them in the mind of the
philosopher only their images.

All science is but the partial reflexion in the
reason of man, of the great all-pervading *reason of
the universe.* And thus the *unity* of science is the

reflexion of the *unity* of nature, and of the *unity* of that supreme reason and intelligence which pervades and rules over nature, and from whence all reason and all science is derived.

§ III.—THE UNIFORMITY OF NATURE.

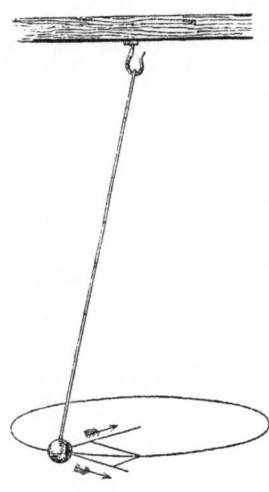

"The harmony of creation is such, that small things constitute a faithful type of greater things."—JEREMIAH HORROCKS, 1637.

Hindrances to science from false analogies.

As real physical analogies form the true ground of inductive speculation, and the power of a right apprehension of them confers that *inductive foresight* which leads to successful discovery, so there are many false views of analogy to be carefully guarded against, involving misconceptions of the relations of physical facts, and leading to mistaken theories and unphilosophical hypotheses, which retard instead of promoting science.

In the earlier stages of induction, laws are assigned of a limited character, circumscribed by many reservations, and qualified by various exceptions and anomalies, real or apparent, yet which must be at least temporarily and provisionally noted as such. Yet it has sometimes happened that such limited views have been converted into positive and general dogmas, from neglecting the obvious caution of always speaking of them *as provisional.*

Exceptions and anomalies maintained as real principles.

Thus sometimes, on the one hand, an unduly limited and restricted view, cautiously entertained by a great philosophical leader, may have been caught up by his followers, misunderstood, and invested with a false character and importance; or, on the other hand, crude ideas may be sometimes hastily thrown out by a great master mind, as first temporary or tentative hypotheses, and then come to be treasured up as absolute dicta by his less discerning disciples, and so have acquired the stamp of permanency, to the great and serious hindrance of scientific progress.

Authority of a name.

Irrationality of dispersion of light.

The once common use of the term " irrationality " of the prismatic spectrum, implied the prevailing prepossession that it might be expected to be " rational," or follow the *same proportion* in all media; whereas, in fact, great difference of ratio prevails. This erroneous first conception long retarded discovery. It, however, shelters itself under the authority of Newton, who assumed it without question, perhaps even as a natural consequence, from the fact, that as spectra given by prisms of all substances are analyses of white light, and can be recompounded into the same white light, the ingredient tints must in all cases be in the same proportion.

Coral reefs.

Nothing could be more marvellous than the hypotheses once universally in vogue as to the formation of *coral reefs*, rising up in so inexplicable a manner from the depths of the ocean; until, by the application of a more correct knowledge of the natural history of the animals, and a simple reference to the common geological phenomena of subsidence, Darwin has divested the whole history of its marvellous character, and restored the equilibrium of inductive uniformity.

When the asteroids were first discovered, an

eminent astronomer maintained that a large planet once moving at their mean distance had *exploded*, of which they were the fragments. Strange as this hypothesis was, it was generally adopted by philosophers, and even calculations were entered into to assign the place at which this marvellous catastrophe took place, and the directions followed by the fragments. But if we simply asked what analogy have we for such an event,—when has a planet ever been known to *burst?* or, indeed, how could such an effect be produced? — its unphilosophical nature would be sufficiently apparent.

Hypothesis of asteroids from the explosion of a planet.

On the other hand, *condensation* of cosmical matter is an hypothesis which has gained ground from many probable analogies; and the supposition of a ring of such matter, out of which these minute bodies (probably existing in vast numbers) have been condensed, formerly thrown out in a public lecture *, has been sanctioned by the authority of Mr. Adams, in an address from the chair of the Astronomical Society.†

To take another instance: the authority of New-

---

* Royal Institution, April 7. 1848: see Athenæum.
† Astr. Soc. Notices, 1853, vol. xiii. 143.

Geometri-
cal method.

ton's name and example undoubtedly, for a long series of years after his death, powerfully influenced for the worse the tendencies of the mathematical spirit of England towards an exclusive preference for the geometrical method : or where necessity might compel the use of analytical processes, still an entire devotion to the letter of Newton's fluxional notation restricted their application, and long continued to make the great advances of the continental analysts a sealed book to the English student, and to retard the progress of investigation in this country.

Unity of
composition
in physi-
ology.

In a similar spirit, without any more disparagement to the great name of Cuvier than to that of Newton, it has been a subject of complaint on the part of a large and increasing school of physiologists, that a too prevalent devotion to the *teleological* methods which he so peculiarly supported, and which derived so undue a preponderance from the authority of his name, has been a great hindrance to the progress of the more extended views opened by the higher principle of " unity of composition " advocated by the school of Geoffry de St. Hilaire, which Cuvier so strenuously opposed, and which the influence of his name, was long so potent unduly to repress.

The prepossession arising from Newton's dictum, Change of refrangi- "To the same ray ever belongs the same refrangi- bility. bility," seems long to have operated as a bar to even imagining any theory which involved an opposite idea. And it was accordingly as a sort of paradox that Professor Stokes ventured to announce his important discovery of a change of refrangibility, which affords the key to so wide a range of curious phenomena of light, including and generalising the singular results before obtained by Sir J. Herschel and others of Sir D. Brewster.

The higher and wider extension of analogy and Slow progress of generalisation is not effected at once and at first. generalisation. The earliest, and sometimes the most highly and justly valued labourers in particular departments and fields of research, as collectors of facts, are not always those best able to perceive the broader connexion of grand principles; and hence are the more apt to cling to such prepossessions as those just alluded to. Even when many classes of facts have been successfully made out, it requires time, and the appearance of some genius of more rare original power, to indicate at once a comprehensive theoretical principle by striking out some general conception, startling, perhaps, from its very simplicity, yet revolutionising the whole science.

Great principles reluctantly adopted.

And even when they are proposed, such more elevated views are not at once appreciated or understood, not merely by the many, but even by the cultivators of science. When Galileo opened the path of all true astronomy by the simple maxim that the same laws of motion which hold good on the surface of the earth apply also throughout the celestial spaces, and when Lyell did the same thing for geology, by maintaining that the analogy of real and existing causes ought to be extended through all the immeasurable periods of past time; — neither was at first admitted without much difficulty and opposition, not so much arising from the mere ignorance of the many, as from the preconceptions of the few.

Dislike of theorising.

Some very eminent men of science have been prone to cherish an intellectual disposition too strongly opposed to all indulgence in hypothesis, and have evinced a very stringent determination to keep to what is regarded as the exactness of demonstrative science, with an especial abhorrence of any thing wearing the appearance of theorising; which they would most carefully shun under the idea of

its being metaphysical and visionary, and altogether
at variance with the severity of all that is worthy
the name of real science, or, at any rate, beyond
their province to pursue; which is surely as much a
fault in the one extreme as the spirit of fanciful
hypothesis is in the other. Yet, men of the most
comprehensive minds are the most ready to admit
the value of such speculative ideas if well formed.

" Beside positive knowledge," says Humboldt,
" stand conjecture and opinion — a philosophical
science of nature strives to rise beyond the limited
requirements of a bare description of nature. It
consists not . . . in the barren accumulation of isolated facts. The curious, the inquiring spirit of
man must be suffered to make excursions, . . . still
to surmise what cannot be positively known." *

We have already observed that all induction implies a *primary adoption* of a *certain amount of hypothesis;* and the secret of its success in any instance
lies mainly in the happy selection of such hypothetical grounds, and not in the mere accumulation of
facts. It is by the peculiar capacity for seizing *sound*

The philosophy of conjecture.

---

* Cosmos, p. 252., 1st trans.

analogies in these first hypotheses that the highest philosophical genius is mainly characterised.

Some persons speak as if all conjecture were alike delusive; but wise and skilful conjectures are very different from hasty and crude guesses; and the comparative probability of several hypotheses, all purely imaginary, admits of many degrees; and to reduce it to something like fixed principles would constitute no unimportant branch of mental science, — the logic of anticipation, the philosophy of the unknown.

Belief from analogy before demonstration.

It was in fact nothing else than the firm persuasion of the truth of great and high principles of philosophical analogy, and the inherent force of sound ideas of probability, which so powerfully influenced those who were the first assertors of the solar system of the planetary world, and even martyrs to its cause, before it had received any absolute proof from the application of mechanical principles; and when the assertion might be called merely conjectural. Yet it was a conjecture of that highest class which is formed by genius in its loftiest moments of inspiration, derived from an enlarged contemplation of the harmony of nature; and, we may add, in like manner it is, that on the other hand the antecedent incredibility of an alleged phenomenon weighs more against it with a sound philosopher, than many assertions in its favour.

A beautiful example of this kind of anticipation is found in the correspondence of the unfortunate but pre-eminently promising Jeremiah Horrocks; when, after objecting to some theories of Kepler, to account for the planetary motions, he adds, " It appears to me, however, that I have fallen upon the true theory, and that it admits of being illustrated by means of natural movements on the surface of the earth, for *nature everywhere acts according to a uniform plan, and the harmony of creation is such that small things constitute a faithful type of greater things.*" *

It was to illustrate this " true theory," that he devised that beautiful experiment (the most instructive which the lecturer can exhibit even at the present day †) of the freely suspended ball revolving in an ellipse under the combined influence of the central and tangential forces ; and in addition showing the motion of the apsides.

Mere conjectural hints at explanation of obscure

Example of Horrocks.

---

* In a letter dated Nov. 23. 1637.   See Grant's Hist. of Astron. p. 425.
† See Vignette.

Conjectural explanations only to show that phenomena are not in their nature inexplicable.

phenomena may in many cases be thrown out, which may seem to some fanciful and idle, and may be easily turned into ridicule; when the real object and meaning is only to show that the phenomena in question are not necessarily of such a nature as to be beyond the boundary of *legitimate investigation,* or *possible explanation.* In such instances, to show a *bare possibility* is all that the case requires; the language sometimes employed is perhaps censured as fanciful and evasive, or as a mere disguise for ignorance; but the real purport of such suggestions is overlooked; they are not put forth as pretending to be *complete explanations;* the point that is aimed at is merely to show that the phenomena in question are not necessarily of a kind *outstanding* and *setting at defiance all physical explanation :* now an event cannot be set down as *inexplicable* to our faculties, so long as any possible or imaginable combination of physical conditions can be suggested as capable of furnishing *a plausible explanation* of it.

Example. Nebular theory.

Thus, for example, the nebular hypothesis of the origin of the planetary system was thrown out by Laplace as confessedly a mere conjecture: yet one which was founded on rational probability; and tended to show that the observed peculiarity of the motions of the planets being all in one direction, was *not absolutely inexplicable* on physical principles, and evinced a probability that on this, or some equivalent principle, the origin of those motions might reasonably be expected to find a solution.

Yet further: to this uniformity in the motions of the system there appears one striking exception in the satellites of Uranus, which are at once *retrograde* and *highly inclined;* as they would be if originally *direct* and then turned over *beyond* the perpendicular. Such a disturbance could not occur from the action of any existing planetary attraction: but, *in the state of nebulosity*, it is far from impossible to conceive some action of the kind among the multitude of conflicting forces then acting.   No sound philosopher doubts that the effect was due to *some* regular cause: the nebular hypothesis may serve to suggest that the conception of such a cause is not wholly beyond the limits of physical analogy.

Anomaly of satellites of Uranus.

The progress of physical discovery may, it is true, be sometimes slow, and the appearance of objections and difficulties so formidable, as to damp the ardour of research, or even to give some colour to the

Difficulties in research not to stop it.

H

insinuations of those who may be sceptical as to the pretensions of philosophical advance, or entertain jealous or hostile feelings towards such pursuits. But the inquirer, truly imbued with inductive principles, will not despond.

Lesser difficulties not to hinder reception of great principles.

There is one grand maxim of pre-eminent value in philosophic inquiry generally, and which finds a peculiar application under such circumstances as those just referred to, viz., that *having once grasped firmly, a great principle, we should be satisfied to leave minor difficulties to await their solution,* assured that in time the progress of discovery will clear them up as certainly as it has now cleared up difficulties once quite as formidable and paradoxes quite as inexplicable. It has been by adherence to this rule that all great philosophical systems have made their way and finally triumphed over error. The Newtonian theory was beset by palpable contradictions in its results till many years after Newton's death; yet all sound philosophers embraced it. The motion of the apsides of the moon's orbit was, with singular honesty, confessed by Newton to be in fact *nearly twice as great* as calculation from theory made it: and this contradiction remained an outstanding palpable ob-

Example. Newtonian system and lunar apsides.

jection, yet without occasioning any misgiving in the mind of sound philosophers as to the general truth of gravitation, till the error was explained, and the calculation rectified by Clairault.*

Up to the present time, *all* the anomalies of the The tides. tides are by no means reduced under the dominion of theory: yet no sound philosopher doubts the truth of the principle that they are due to the solar and lunar attractions.

The theory of gravitation, again, was really de- Motion of Uranus. fective up to the present day: the motions of the planet Uranus, as calculated by theory, were found to be every year becoming more and more discordant with observation, and theory was completely at fault: until the calculation of Adams and Le Verrier showed that the anomalies could be explained by the supposition of the disturbance occasioned by an exterior planet moving in a certain orbit; and, at the time, at a certain point in that orbit; as was directly verified by the observations of Galle and other astronomers.

So again, the undulatory theory of light now un- The wave theory of light.

---

* Princip. bk. i. sect. ix. prop. 45. cor. 2.

hesitatingly accepted by all mathematicians is still
confessedly defective in its application to several
phenomena, especially the more extreme cases of
prismatic dispersion.

Melloni's
anomaly in
the solar
rays.

The rays of the sun under ordinary circumstances
possess a heating power in proportion to their intensity.   When analysed, though the heating power
differs greatly for different primary rays, and is not
proportional to their illuminating power, yet it
nowhere exists without rays which may be rendered
visible, nor is any visible ray destitute of such power:
and for the same ray under the same conditions the
heating is proportional to the illuminating intensity.

The alleged invisible heating rays discovered by
Sir W. Herschel beyond the red end of the spectrum,
are not a real exception, since by proper precautions,
such as using a deep blue glass, red light may be
rendered visible far beyond the ordinarily seen
boundary.

To this general law one, and one only, outstanding exception occurs in an experiment recorded by
Melloni, viz., that, with a certain green glass, the rays
transmitted when concentrated by a lens, are intensely bright, but totally destitute of heat.   This

is a solitary exception — a breach of all analogy — unsupported by any corroborative experiments : *and as yet unexamined* by any critical experimenter. It is then simply an anomaly *provisionally*.

The anomaly that water is at its greatest density at about 40° F., and below that *expands* with decrease of temperature, has been held by some to be a marvellous and peculiar outstanding fact, setting all theory at defiance. Maximum density of water.

Yet no truly inductive philosopher for a moment doubts that it is really a part and consequence of some higher law of which the ordinary law of expansion is a part.

Indeed, Berthollet speculated on the subject, so far, at least, as to maintain that the cause, whatever it be, which produces crystallisation, is in operation in expanding the water before the crystals of ice are actually formed, and which are specifically much lighter than the water. He even states it as a general law that " the causes which determine the changes of constitution of bodies exercise an action, the effects of which are evident before the changes of constitution have taken place." And this property

in water is not altogether an anomaly as compared
with what takes places in antimony, iron, and bismuth.

Instances also occur in certain substances in a
fluid state which *instantly* solidify on the application
of an extraneous body — proving that the particles
are already in *a state of strain,* and require only some
extraneous agency to bring about that change in
their arrangement.

Other suggestions of a theoretical kind have also
been made : but at any rate, we see enough to show.
that the phenomenon is *not one of necessity outstanding all explanation,* and that it very probably
will ere long be brought under the dominion of
theory.

Principle of
uniformity
throughout
nature the
essence of
all induc-
tion.

The system of inductive reasoning in its full extent centers in the *conviction of the universal and
permanent uniformity of nature.* This, as was observed at first, has been emphatically and truly called
" *the inductive principle.*" It is this which points to
the great archetype of UNITY ; to which all our subsequent conclusions minister increasing confirmation ;
and from the influence of such a first principle in
our inquiries arises all that distinguishes true science
from mere empiricism, and an elevated philosophy

from the grovelling and mechanical accumulation of mere millions of facts.

And we may remark that this idea, in its proper extent, is by no means one of popular acceptance or natural growth.   Just so far as the daily experience of every one goes, so far indeed he comes to embrace a certain persuasion of this kind ; but merely to this limited extent, that what is going on around him at present, in his own narrow sphere of observation, will go on in like manner in future.   The peasant believes that the sun which rose to-day will rise again to-morrow; that the seed put into the ground will be followed in due time by the harvest this year as it was last year, and the like; but has no notion of such inferences in subjects beyond his immediate observation.

Not an intuitive or natural belief.

And it should be observed that each class of persons, in admitting this belief within the limited range of their own experience, though they doubt or deny it in everything beyond, are, in fact, bearing unconscious testimony to its universal truth.   Nor, again, is it only among the *most* ignorant that this limitation is put upon the truth.   There is a very general propensity to believe that everything beyond com-

Commonly doubted or denied beyond narrow limits.

mon experience, or expressly ascertained laws of
nature, is left to the dominion of chance or fate or
arbitrary intervention; and even to object to any
attempted explanation by physical causes, if conjec-
turally thrown out for an apparently unaccountable
phenomenon.

Acquired
only by
philosophi-
cal study.

The precise doctrine of the *generalisation* of this
idea of the uniformity of nature, so far from being
obvious, natural, or intuitive, is utterly beyond the
attainment of the many. In all the extent of its universality it is characteristic of the philosopher. It is
clearly the result of scientific cultivation and training, and by no means the spontaneous offspring of
any primary principle naturally inherent in the mind,
as some seem to believe. It is no mere vague persuasion taken up without examination as a common
prepossession to which we are always accustomed ; on
the contrary, all common prejudices and associations
are against it. It is pre-eminently *an acquired idea.*
It is not attained without deep study and reflection.
The best informed philosopher is the man who most
firmly believes it, even in opposition to received
notions; its acceptance depends on the extent and
profoundness of his inductive studies.

Throughout the range which science opens to us we find the several classes and orders of phenomena defined by laws of increasing generality, and thus intimately connected and bound together, so that every part is essential to the coherence and unity of the whole. But when we have arrived at the highest of such generalisations to which science has yet attained, those most comprehensive laws, in the *existing* state of our knowledge, seem diverse, disunited, and not as yet connected by any common higher principle; yet we cannot for a moment suppose this to be anything else than the result of our ignorance; they must each be really subordinate members of some greater group. Future research will undoubtedly connect them together by a common principle, of which at present we can form no more conception than the predecessors of Newton did of universal gravitation, or than he did of electro-magnetism, or geological epochs. Discoveries are being made every day; and the very next important physical discovery will as assuredly effect an union between some two or more classes of phenomena at present not so connected, as the last discovery has done. New *phenomena* are being con-

Increasing conviction by more extended research.

tinually detected. Not more surely does this happen than it is sooner or later followed by the disclosure of more comprehensive *laws*. The progress of discovery is as certain as the extent of nature is unlimited; and the subordination of *species* of phenomena to *genera*; of these genera, again, to *classes* or higher genera, and so on, must be as unbounded as the succession of phenomena.

Idea of chance.

The universality of law and order is the distinguishing conviction of the inductive philosopher; by this, in fact, *science* is elevated into *philosophy*. One main test of its force and extent is the exclusion, in consequence of its admission, of the very *notion of chance*, or of the possibility of any events in the universe really happening *at random*. In fact, the very term chance implies a theory; and if we would examine its meaning, and employ it in a strict sense, we should find that what we really mean can never properly amount to more than a confession of *our ignorance* of the mode or order in which certain events have taken place. If we take any portion of the natural world, or any class of phenomena of which we know least, and which appear most fortuitous, can we correctly say more than that

we are ignorant of the laws by which it is regulated?
Yet, while in saying that any phenomena appear
capricious or fortuitous, we simply admit our ignorance of the laws by which they are governed, no
inductive philosopher for a moment doubts that they
are regulated by *some* laws.

To take an example: the apparently irregular
mode of distribution of the fixed stars through the
heavens, might seem, at first sight, to justify the
belief that their arrangement and position in the
universe might be wholly *fortuitous*, and such probably may be the idea in uninstructed minds, and
they may perhaps feel disposed to ridicule the somewhat bold but characteristic idea of Des Cartes *,
who says that he meditates an attempt to investigate
the *cause* of the position of every fixed star. Yet
the very fact that these masses at least have the
property of transmitting light to us, and consist of
matter of some kind, and have been in some instances
proved to be subject to the law of gravitation, in-
stantaneously asserts for them a sort of claim of
kindred with matter around us and with ourselves

Example.
Distribu-
tion of fixed
stars.

---

* Epist. 67.

and dispels every shadow of doubt that they are disposed according to *some* physical law, under the influence of *some* determinate physical forces.

Struve's re-
searches.

When we come to examine the masterly and profound researches of Struve (disclosed in the " Etudes d'Astronomie Stellaire ") we begin to feel more convinced that even in the seemingly capricious distribution of these remote masses through the abysses of space, we obtain a glimpse of *order*, if only from the mere fact that by the consideration of *averages*, some sort of classification is effected ; and from that happy combination of arguments brought together from such various sources, which none but an inductive genius of the highest order could have planned, and nothing but consummate mathematical skill could have worked out, conclusions of high generality and profound interest are elicited in a subject, at first sight, seeming to baffle inquiry.

Geological
elevations.

Nothing would appear, at first sight, more devoid of all order, or apparently fortuitous, than the directions assumed by those elevations and fractures of strata which diversify the surface of the earth with mountains and valleys, precipices and plains. Yet the accurate observations of geologists, combined with the theoretical indications of dynamical science, have even now begun to throw some light on the probable laws of these seemingly arbitrary manifestations of power, and to connect them with the all-pervading principles of regularity; and though we may not be disposed to assent to the precise theory of Elie de Beaumont, yet it at least gave a right direction to inquiry, and the exact deductions of Mr. Hopkins, place the general dynamical principle of lines of upheaval beyond question: and leave no doubt that a comprehensive mechanical theory will eventually be worked out, and the most monstrous geological "catastrophes" reduced to order and system.

Among the ancients we know the several forms of belief in blind fate or chance were not merely popular delusions, but deliberate persuasions, which divided philosophical sects: the advocates of the fixed necessity and eternal destiny of the world, and the supporters of the Epicurean doctrine of the formation of the material universe out of a fortuitous concourse of atoms.   But in the age and under the influence of the inductive philosophy, no such dreams can for a moment obtrude themselves.   The definite and positive spirit of this system strikes at the root of such

Fate and chance excluded.

vague and unmeaning expressions — the mere dis-
guises of human ignorance.   It demands *what* chance
and fate are.   It appeals to the great principle of
uniformity, and the regularity of physical causes;
and feels warranted in affirming that in all cases,
however incapable they may at first appear of reduction to any kind of system, there yet *must be*
in reality as perfect, though to us unknown, observance of determinate laws in their production,
as in any cases we are most familiar with.   Profoundly *adjusted order* is utterly inconsistent with
blind destiny, mechanical causes with chance.

No limits to the application of induction.
It is the proper business of inductive science to
analyse whatever comes before it.   We cannot say
that any physical subject proposed is incapable of
such analysis, or not a proper subject for it, until it
has been tried and found to fail; and even then, the
result is not unprofitable; — we know the precise
point at which the failure has taken place, and the
exact cause of its occurrence.   It is a main characteristic of sound philosophy, that it draws the line
precisely between the known and the unknown;
and teaches us not only why we understand the
one part, but why we do not understand the other.
Yet the unknown regions on the frontier of science

enjoy at least a twilight from its illumination, and are
still brightened by the rays of present conjecture,
and the hope of future discovery. We can never say
that we have arrived at such a boundary as shall
place an *impassable* limit to all future advance,
provided the attempts at such advance be always
made in a strictly inductive spirit. To the truly
inductive philosopher, the notion of limit to inquiry
is no more real than the mirage which seems to
bound the edge of the desert, yet through which
the traveller will continue his march to-morrow, as
uninterruptedly as to-day over the plain.*

When the inductive inquirer finds himself involved
in some great apparent difficulty, and among phenomena which no existing resources of science are
able to explain, which appear to stand forth as irre-
ducible anomalies, and to baffle all attempts at explanation ; however hopeless the problem may seem,
he can never really suppose the case to be *in its own
nature incapable* of analysis, or that the mass of facts
is not really reducible to *some* principles of order,
analogy, and causation, — to the dominion of laws as

Limits of our present knowledge distinct from those of nature.

---

* See note at the end of the section.

yet indeed unknown, and of causes not as yet conjectured, yet as perfectly regular and strictly harmonious as those which govern the most common daily
occurrences, — the fall of a stone, or the ascent of
vapour.

No *real* in-
terruptions
in the order
of nature.
A real break in the connexion and continuity of
physical causes cannot exist in the nature of things.
If such breaks often appear, they are due solely to
our ignorance. Every advance tends to fill them
up; and indeed each physical discovery is nothing
else than an extension of the evidence of continuity,
a fresh link in the connexion of phenomena into óne
consistent whole.

Anomalies
only appa-
rent.
There is no such thing as any class of phenomena
really standing out isolated from all others unconnected by any analogous principle, and truly anomalous in regard to the rest of nature. Yet every
class of phenomena has at some time seemed so;
but it is an illusion in whatever instance it may
now seem to be the case; and one which time will
assuredly clear away, as it has already done so
many similar or greater illusions.

Anomalies
referred to
laws as yet
unknown.
In all apparent anomalies, the inductive philosopher will fall back on the primary maxim, that it

is always *more probable that events of an unaccountable and marvellous character are parts of some great fixed order of causes unknown to us, than that any real interruption occurs.* And further, what may *now* appear the most mysterious, and at present least understood, will yet hereafter be explained by the future extension of discovery.

It may, indeed, be difficult or impossible to apply these considerations *in detail,* and to suggest particular interpretations in subordination to these paramount principles; yet this will not invalidate their general truth: nor need it lead us into extravagant and gratuitous speculations to bring about a precise explanation for which the circumstances do not furnish sufficient data. A truly *rational* inquirer will be content *to let such difficulties await their solution:* and, so far from always seeking such explanations in precise theories, he will admit, on the contrary, that too minute a solicitude to refer every case to KNOWN *causes,* may tend to keep out of sight the broader principle that they may be referable *to some causes as yet* UNKNOWN, but still parts of the same universal order; and may even lead to the disparagement of

I

that principle when, in any instance, such more particular mode of explanation is found to fail.

Instances.   For example : in the present state of science, of
all subjects, that on which we know least is, perhaps,
the connexion of our bodily and mental nature,
the action of the one on the other, and all the vast
range of sensations, sympathies, and influences in
which those effects are displayed, and of which we
have sometimes such extraordinary manifestations in
peculiar states of excited cerebral or nervous action,
somnambulism, spectral impressions, the phenomena
of suspended animation, double consciousness, and
the like.*   In such cases science has not yet advanced to any generalisations ; results only are presented, which have not as yet been traced to laws.
Yet no inductive inquirer for a moment doubts that
these classes of phenomena are all really connected
by *some great principles of order.*

If, then, some peculiar manifestations should have
appeared of a more extraordinary character, still less
apparently reducible to any *known* principles, it
could not be doubted by any philosophic mind that

---

* The reader is referred to "Letters on the Truths contained in Popular
Superstitions," by the late Herbert Mayo, M.D. 1849.

they were in reality harmonious and conspiring parts
of some higher series of causes as yet undiscovered.

The most formidable outstanding apparent ano- Anomalies
will be
malies will at some future time undoubtedly be cleared up
by future
found to merge in great and harmonious laws, discovery.
the connexion will be fully made out, and the
claims of order, continuity, and analogy, eventually
vindicated.

Inductive philosophy has within itself a pro-
phetical warrant to foresee that a time will come
when those things which seem most obscure will
become clear. The well-known prediction of such a
disclosure in the case of the celestial motions uttered
long ago by Seneca*, and fulfilled in Newton, is not
less applicable at the present time, and points to
equally grand openings in all branches of physical
science, which will as assuredly be made at other
future epochs of scientific revelation.

When we arrive at any such seeming boundary of No commencement
present investigation, still this brings us to no *new* of a new
order of
world in which a different order of things prevails; things.

---

* " Veniet tempus, quo ista quæ nunc latent, in lucem dies extrahat,
et longioris ævi diligentia: ad inquisitionem tantorum ætas una non
sufficit; veniet tempus quo posteri nostri tam aperta nos nescisse mi-
rentur."— *Nat. Quæst.*, viii. 25.

it merely points to what will assuredly be a fresh starting point for future research. It is an unwarrantable presumption to assert, that at a mere point of difficulty or obscurity we have reached the boundary of the dominion of physical law, and must suppose all beyond to be arbitrary and inscrutable to our faculties. It is the mere refuge and confession of ignorance and indolence to imagine special interruptions, and to abandon reason for mysticism.

Conclusion.   The consideration of the uniformity of nature leads directly to a more precise — a higher — view of the same great conclusion to which we before adverted generally.

Evidence of a supreme mind.   All induction begins and ends in the conception of order, arrangement, and uniformity throughout nature; and this, however inadequately comprehended by our science, is again the evidence of supreme mind, and the universality of order in time and space, the manifestation of the universality and eternity of that supreme mind.

It has been eloquently observed, " Humboldt thought he could show why and how this world, and the universe itself, is a *kosmos*, — a divine whole of life and intellect, — namely, by its all-pervading *eternal laws.* Law is the supreme rule of the universe; and that law is wisdom, is intellect, is reason, whether viewed in the formation of planetary systems, or in the organisation of the worm."*

And in a similar spirit Œrsted has said: " The progress of discovery continually produces fresh evidence that Nature acts according to eternal laws, and that these laws are constituted as the mandates of an infinite perfect reason; so that the friend of Nature lives in a constant rational contemplation of the Omnipresent Divinity." † . . .

" The laws of Nature are the thoughts of Nature; and these are the thoughts of God." ‡

---

* Chevalier Bunsen's reply to the President's Address, on delivering the medal to Humboldt, Royal Society Anniversary, 1852.
† Soul in Nature, p. 196.   ‡ Ib. p. 20.

NOTE TO PAGE 111.

Of the Baconian philosophy, it has been said by a masterly writer, "It is a philosophy which never rests; which has never attained, which is never perfect. Its law is progress. A point which yesterday was invisible is its goal to-day, and will be its starting post to-morrow." (Macaulay's Essay on Lord Bacon, p. 113. small ed.) But while I cannot refrain from citing this brilliant sentence with all the admiration it deserves (and indeed many others in the same essay are not less worthy of admiration), I feel bound to express my dissent from the *exclusively* practical view which the author takes of the objects of inductive science, and must regard it as hardly less than a profanation of the name of Bacon, to associate it with such unmixed utilitarianism as he would represent to be its aim. At the same time it is right to add, in some other passages the author seems disposed to modify the strong assertion of this view, especially p. 116, where he fully admits the high moral influence and objects of the Baconian philosophy.

§ IV. — THE THEORY OF CAUSATION.

1666.

Vel huic philosophandi modo, vel *veriori alicui.*
NEWTON, *Pref. in Princip.*

Desire to
seek causes.

AMONG our various intellectual propensities, there
is none more powerful or more seductive than the
desire to penetrate into the *causes* of things. We
perceive events going on or results produced in the
natural world; we recognise a number of different
powers or agents at work; and to these, under the

name of *causes,* or, more strictly, *physical,* or, according to some, *secondary causes,* men are prone in imagination to ascribe a sort of energetic *power,* or a coercive *efficiency,* by virtue of which these physical agents produce, or bring about, certain results : a species of active influence by which matter is imagined to act upon matter, and produce a different state of things, in a way exactly analogous to, if not identical with, that in which a voluntary agent exercises his volition on material objects within his control; and thus there is supposed to exist a relation of a peculiar and intimate, yet hidden and unknown kind, not to be traced by our faculties or further explained, yet the essential condition of all real philosophic investigation; and views more or less similar to these seem to have been very generally, entertained among philosophers in former times.

Natural to imagine efficient power.

But when Hume, in his essay on Necessary Connexion, showed that of the existence of this kind of mysterious influence or imaginary power, there neither was, nor could be, *any evidence;* — that in physical events all we could really infer was the mere fact of the *invariable sequence* of the one event called the *effect, after* the other called the cause, — a doctrine so

But unphilosophical.

opposed to the favourite mysticism, which delights in
investing scientific truth in a veil of abstruseness, and
will not condescend to acknowledge any thing intelligible to be true philosophy, was of course not
received without much open hostility from some
parties ; while from others, who felt constrained to
acknowledge the strictness of the conclusion, it

Objections
to simple
view.

obtained a reluctant and modified acceptance. It
was complained of as a meagre, empty, unsatisfactory doctrine, tending to degrade philosophic
speculation to mere matter of fact, and not penetrating below the surface. Thus Lord Kames,
though admitting that no connexion of cause and
effect is discoverable by *reason*, yet contended that
it nevertheless really exists ; for we *feel* and acknowledge that every effect implies a cause, and that
nothing can begin to exist without a cause of its
existence.* That men are prone to *feel* and acknowledge such a notion is perfectly true ; but the very
question at issue is, do they do so *correctly*, or on any
real philosophic ground ?

Without here pretending to go into the various

---

* See Burton's Life of Hume, i. 427.

discussions of the subject which have taken place among subsequent philosophers, and disclaiming all controversy, I will merely remark that at the present day the question is still kept up by advocates of each extreme — the one party contending for the old idea of efficient causation and necessary connexion, and the other adopting the view of Hume, modified by one or two qualifications, yet maintaining the principle of a simple, invariable (or as Mr. Mill terms it, " unconditional ") *sequence of events;* and agreeing therein with the French school of positive philosophy, as expounded by M. Comte, in totally rejecting the idea of causation in physical phenomena, in the sense of *efficient* power, as a notion wholly beyond our capacities to define or reason upon, and therefore unphilosophical.

Different opinions at the present day.

My own views of the subject have been expressed in a work published long ago *; but it may be desirable to offer some further explanation of them, after a careful examination of what has been advanced since, whether in support and elucidation of

General view of the case.

---

* The Connexion of Natural and Divine Truth, London, 1838. See also my Essay on Necessary and Contingent Truth, Oxford, Ashmolean Memoirs, 1849.

the great step made by Hume, or in attempting a
retrograde movement and a revival of the exploded,
but naturally popular notion, of efficient agency; the
personification of matter and mechanical forces, derived from an imaginary analogy between physical
action and that of voluntary agents. I conceive that
all real philosophical analysis of the case must end
in an entire repudiation of such fanciful notions, involving as they appear to me to do, a confusion of
ideas, which I think may be completely avoided
by the simple distinction between *physical* causation,
or the action of *matter* on *matter*, and *moral* causation, or the action of *mind* on *matter*.

> Distinction of physical and moral causation.

To take the simplest example: I throw a stone,
which brings down a bird; my volition is said to be
the *cause* of the stone's flight; the impact of the stone
the *cause* of the bird's fall. The word " cause " is
here used in two totally different senses: in the first
instance, signifying *moral*; in the latter, *physical*
causation. Rejecting altogether the idea of efficient
causation, as wholly inapplicable in relation to physical effects, however pleasing to the fancy, I conceive that the true theory of physical causation
includes the simple idea of an invariable or " uncon-

> Physical cause implies sequence in *relation*.

ditional" sequence of facts (meaning sequence in *relation* not necessarily in *time*); yet I contend that there is implied also a *connexion, not* in the *events* in the way of *physical agency,* but in the *reason* and *logical dependence* of the two ideas.   The phenomenon or property assigned as the cause or antecedent has undoubtedly a *necessary connexion* with the effect or consequent, when it supplies the *explanation* of it: when the latter is a *consequence in reason* and *theory* from the former — when, in a word, the cause is a *more general* and *better understood class* or *genus* of phenomena to which we can refer the effect, as a particular *species.* Necessary connexion, *in reason* not in the events.

For example: *friction is the cause of retardation* of motion.   There is a mere sequence of two phenomena.   Yet there is also a necessary connexion between them, though not in the sense of efficient power; for we conceive the notion of friction, and we then *reason from it,* that retardation will be a *necessary consequence.*   But there are many cases where this kind of connexion is less strong and instructive.   *Friction is the cause of heat;* but we do not know enough of the nature of friction to be quite certain *why* or *how* it produces heat, Example: Friction and retardation. Friction and heat.

though we may *conjecture* it to a certain extent. Here, then, the connexion is not so necessary. Again, *friction* (in certain bodies) *evolves electricity.* Here we have still less of connexion ; there is only a *sequence.* In other words, *physical causation admits of degrees.* But this kind of *connexion* in *reason,* even its highest degree, is totally remote from any analogy with *moral* causation, or the sense of power or effort in a voluntary agent.

Cases are sometimes alleged of particular incidental events which are the immediate means, or instruments, or occasions for other events taking place, and are thence called their *causes :* as the opening of floodgates is said to be the *cause* of the flow of water. Yet it is urged gravitation or pressure might, with equal or greater truth, be called the cause; that is, we here use the word cause in a more limited sense. When we speak of physical causes in a philosophical sense, we must recur to the idea not of *mere* sequence of events, but of sequence in *reason.* The pressure of the fluid is doubtless the physical cause of its overflowing: the particular case of floodgates is only an incidental occasion for its action ; only a particular form of the more general ac-

Friction
and electricity.

Fallacies
to be
avoided.
" Cause "
sometimes
used for
" occasion."

Sequence
in time.

tion of pressure. Some writers, again, fall into, or per-
plex themselves with, what is nothing more than the
old fallacy of " post hoc ergo propter hoc ; " mistak-
ing a succession *in time* for a succession in *reason
relation ;* as in objecting that we thus make day the
cause of night, and the like.

In fact, the circumstance of *time is wholly irrelevant* Order of
time irre-
to the idea of cause and effect. We may convince levant.
ourselves of this by referring to the numerous instances where the phenomena are *cotemporaneous.*

Thus the *pressure* and the *density* of elastic fluids Cause and
effect often
are *cotemporaneous* conditions : yet the first is the coexistent.
cause of the second. Evolution of heat with condensation, and absorption of it with expansion, are
coexistent. Chemical decomposition in the elements
of a galvanic battery, and the production of the
galvanic current, are *simultaneous.* In these and Connexion
in reason
many similar cases, then, of cause and effect, there is admits of
degrees.
no sequence at all in time. The question is as to a
sequence in reason, and this admits of *many degrees,
according to the higher degree of generalisation implied.*

In the last instance of chemical action and galvanism, the effects are *not only simultaneous,* but
also *convertible.* Chemical action is the *cause* of

galvanism; and galvanism is also the *cause* of chemical action.

This, however, is no contradiction or confusion of ideas : it depends simply on *the relation in which we view the case.* What is meant is, that chemical action, *in the instance of the galvanic battery,* is the cause of the galvanic current; and again, the galvanic current, *in the instance of an experiment performed by that battery,* is the cause of chemical decomposition. We are speaking of different cases.

In some cases cause and effect relative terms.

Thus, in these instances, the use of the terms cause and effect is *relative* to the circumstances and conditions which we are at the time supposing.

Sometimes convertible.

Or again, it is said, " magnetism is the cause of electricity, and electricity is the cause of magnetism; " but what is meant is, that in certain experiments, magnetism is so applied as to produce electricity; and in certain *others,* electricity is so applied as to produce magnetism. They are not cause and effect convertibly *in the same sense,* or under the same circumstances. We view them as thus convertible *in different relations.*

Mr. Grove* has considered these cases, and has

* See Correlation of Forces, p. 6.

been led to the conclusion that, " abstract secondary causation does not exist," or, in other words, cause and effect are purely relative terms, *in such cases as those he has considered,* which is exactly what has here been shown.

This is equally true for the other cases which form the subject of Mr. Grove's valuable discussion. The mutual actions of all the imponderable agents, he shows, *are correlative,* or convertible into each other, but no one the essential cause of the other. They are so in different points of view, or on different grounds of relation, as just explained.

But again, as to *the nature of the connexion* between the facts in either case: In the instance of pressure and density of elastic fluid, we perceive *a necessary connexion in reason;* by abstract mathematical reasoning we can infer the one from the other, starting with a definition of an elastic fluid. In the instance of galvanism and chemical action, *we know less* of the connexion, and perhaps cannot show *abstractedly why* one *must* accompany the other.

In these cases a higher connexion in reason.

In the same way in the mutual actions of the other imponderable agents, we cannot reason *abstractedly* to the effects. Where no relation in

reason is yet made out, we can only recur to the
mere law as experimentally established, and traceable
to no higher principle.   In such a case, either phenomenon may be cause or effect relatively to the
other, as seen under different points of view.

In higher
cases no
converti-
bility.

But in other cases where we have attained a
higher and more satisfactory view of a *connexion in
reason*, physical causation is more substantially determined.   When we can ascend to an abstract
principle, and reason conclusively from that principle, that such a result *must* take place as a *consequence* of it, we assign a positive and fixed physical
cause in that principle to which we refer ; and we
cannot reverse the order of relation.

We could not speak of (*e. g.*) gravitation and the
tides as cause and effect to each other *convertibly :*
or of the connexion of ethereal vibrations and
periodical colours as *relative* or *interchangeable.*
Whenever we *can* thus mutually convert causes
into effects, it only shows the little advance yet
made in theoretic generalisation in that particular
subject.   While, again, in regard to the particular
cases of the imponderable agents just considered, it
is extremely probable that future discovery will show

them to be all merely different modifications of one common principle, and thus easily capable of 'convertibility in their effects.

We thus place the theory of causes in its proper relation to that of inductive laws. In assigning physical causes, we refer a particular phenomenon to a more general, — we refer an *event* to a *law;* and the more strictly we analyse our conceptions the more clearly does it appear that we can never arrive at, or, need require, any higher or more intimate connexion than that of successively higher generalisation; by virtue of which to trace a real and satisfactory relation between physical phenomena and the higher abstract principles which combine them together by a "necessary connexion" of *reason,* as parts of a great harmonious whole.

Relation of causation to inductive laws.

Yet against this view, it is urged that it is *unsatisfactory;* that the mind still *craves* a more intimate sense of the connexion of events; and that the universal opinion, and common sense of mankind, rejects such cold and dry abstractions, and naturally adopts the more congenial belief in "efficient" causes, and active power in bringing about physical phenomena. This, however, is nothing more than

Prejudices opposed to this view.

K

an instance of the general reluctance with which the untaught mind adopts any strict philosophical conclusions: all exact analysis of physical phenomena seems cold and distasteful to the unpractised conceptions, and as men soon dislike what they cannot easily understand, this doubtless is often the origin of the vulgar prejudice and hostility against the higher views of science, and the spirit of abstract philosophy.

General belief no proof.

But even were this persuasion as to efficient causation really *universal*, were it not in fact opposed by as large a section of philosophers as those who uphold it, still, *universal belief* would be no proof of its truth. All mankind, three centuries ago, had a universal belief in the geocentric system. Such general persuasion, if anything, would rather suggest a caution that the popular notion may be a popular delusion. In this, as in another sense, we may say, " argumentum pessimi turba est." And doubtless nothing is more difficult to the unphilosophical mind than to be satisfied with *negation* : to learn the humiliating lesson of its own ignorance.

Idea of efficient cause supposed natural.

Some writers have dwelt upon the idea of causation as arising out of some fundamental principle in the constitution of our minds, and have enlarged on the relation of cause and effect as one under which we are constrained to arrange our perceptions, just as the nature of a machine determines the changes of matter subjected to its action.

But all this is beside the present question, which refers to *what is* the relation in question; and the natural proneness, or necessity, if it be so, is nothing more than a disposition to create in imagination a kind of connexion which does not exist, and to overlook the real and simple relation in which the necessity is simply a necessity of *logical* sequence, applied to a sequence or relation of facts.

The notion of efficient causes is doubtless captivating to the imagination as seeming to let us more intimately into the secrets of Nature. Yet it must be sternly rejected by those philosophers who would adhere strictly to the cautious and positive spirit of the Baconian induction. *But delusive:*

In fact, there is an inherent inconsistency in such an appeal to efficient causation. For if this mysterious idea be that which alone supplies a satisfactory insight into the mechanism of the natural world, it must follow, that of the real causes of phenomena we *and involves inconsistencies.*

know nothing, even in the cases supposed to be most fully and satisfactorily established : *e. g.* if anywhere, surely in the principle of gravitation we must acknowledge a cause which furnishes a complete explanation of the planetary motions ; yet the nature of gravity as an efficient cause is confessedly wholly unknown. To the advocates of this view, therefore, the theory of gravitation must be wholly unsatisfactory, and we cannot be said to have attained any real knowledge of the cause of the celestial motions.

Yet it has been urged by those of this school that the notion of a mere sequence is utterly insufficient, that it is little to say such a phenomenon is produced by virtue of such a *law*, — that a law of action is not action; and the like : nor does the mere reference to a bare *sequence* of events afford any very substantial answer to the objection. The view, however, above explained seems to remove such difficulties. To refer any class of phenomena to a higher *genus* is really to explain its nature: to assign such a governing principle is to show on what the phenomenon depends in the connexion of reasoning, which is the only real idea of its necessary relation to a cause.

Logical connexion of sequence sufficient.

This view of causation is in the closest conformity to the grand idea of *unity* pervading the order of physical things, and at the same time banishes all those partial and extraneous suppositions which tend to disparage and mar that grand conception, by the introduction of the obsolete and, in fact, unintelligible notions of efficient causation and active power in physical agents;—the chimæras of an older school and a past age; though attempts are being continually made to revive them.

It follows, from the view thus taken, that there is no contradiction or absurdity (as there must appear to be on any conception of efficient causation) in the assertion that " causation admits of *degrees*," and this *in itself*, and not merely in the extent to which we apprehend it.

If the true notion of *cause* be that of referring the more limited phenomenon or law to the higher or more generalised principle, then it is clear that this relation is *really more complete* and *intimate*, in proportion as such fact is referred to a successively higher or more comprehensive law or principle. We view every phenomenon as connected not with one cause, but with a series of causes rising one above another in generality, and evincing a more intimate and satisfactory relation in proportion as they rise in the scale.*

Cause not power: nor origin.

The view of the question which I have followed, and the rejection of the idea of *cause* in the sense of *power*, as founded on any strict inductive principles, is also important in its consequences as also removing from inductive philosophy all notion of *cause* in the sense of *origin*. The absence of any *essential* relation of sequence *in order of time*, between an effect and its physical cause, in fact excludes the idea of a physical cause pre-existing in time, and *producing*, or *giving origin* to, the existence of another object. The celebrated dogma " nothing can exist without a cause," according to this view is wholly unmeaning, and destitute equally of foundation and of application. There are, of course, innumerable cases in which we can trace the existence of a particular body or being to the operation of certain physical agents, or causes; as the formation of a chemical compound from the

* I have in this 2nd edition added this paragraph with special reference to an objection raised by a candid and able critic against the idea of causation admitting of *degrees*, which he supposes can only be in regard to our more or less perfect knowledge. This I have, I trust, now clearly pointed out, is not the case; the objection, I think, arises from a lingering and unconscious adherence to the old *notion* of causation, along with the retention of the *term*. See above, p. 124.

action of certain affinities; the production of a plant or an animal from a seed,—an ovum,—a parent; and the like: but these are not *originations*, but *changes*, and do not strictly exemplify the maxim at all. The only intelligible sense of such a proposition is that *everything is traceable* (actually, or probably) *to some higher principle:* in which sense, of course, I fully recognise it.

According to the old theory of efficient causes, a species of active power is imagined to reside in natural agents, or to act through them, which constitutes the alleged necessary connexion of physical effects with their causes. This is always affirmed to be something of a nature not at all cognisable by our faculties, and dependent on conditions of an occult and mysterious kind. *Efficient causation leads to fanciful physical theories.*

Hence it seems to be supposed that anomalous deviations occasionally arise, and the idea of efficient causes is specially favoured by those who are fond of imagining marvellous influences of a kind, distinct from, and even interrupting, the ordinary course of natural events. Such, we must suppose, are the catastrophes and convulsions of nature—failures in creation—random scatterings of matter, and other like

notions which are sometimes resorted to as a con-
solation to the wearied theorist when matter-of-fact
inferences seem for a moment to have reached their
limit.

Opposed to immutability of physical order.

Such ideas, however, are not only delusive in
themselves, but are radically opposed to the grand
truth of the uniformity of nature, the unity of
arrangement and design, and by consequence *so far*
would tend to impugn the evidence of higher truths.

Confusion of second causes and First Cause.

Yet we hear the notion of " efficient causation " in
nature upheld by some as of a peculiarly religious
tendency ; while (with strange inconsistency) in
popular estimation the study of " secondary causes "
is accused of being hostile to the belief in a " First
Cause." And (from the same confusion of ideas)
the denial of efficient causes, and the assertion of a
mere sequence of phenomena and laws, is charged
with having the same dangerous tendency even in a
higher degree.

Charge against the Newtonian system.

Thus, Leibnitz brought against the Newtonian
philosophy the strange accusation, " that it deserts
mechanical causes, and is built upon miracles, and
recurs to occult qualities."*

---

* See Edleston's Correspondence of Newton, p. 153.

It seems to have been under the belief of this singular charge that Pope originally wrote the well-known lines which appear in the earlier editions of the " Dunciad," —

> " Philosophy that reached the heavens before,
>   Shrinks to her hidden cause, and is no more ; "

which, had the fact been as supposed, would have conveyed as perfectly just a censure in the second line, as it does the characteristic of a *true* philosophy in the first, as *leading to*, not *starting from*, the belief in a Deity.

Whereas, when undeceived as to the fact, the lines which he substituted in the later editions,—

> " Philosophy that leaned on Heaven before,
>   Shrinks to her second cause, and is no more,"

embody the whole vulgar misconception and confusion of ideas respecting First and Second Causes, while they are, in any sense, as wholly inapplicable to the *Newtonian* philosophy, as the former.

A recurrence to (what is at least) the simple and intelligible view above expounded, would remove altogether the whole mass of difficulty, confusion, and objection, in which we are thus entangled, and which is involved in the notion, so commonly alleged, of an eternal succession of secondary causes, ex-

Difficulties removed.

cluding the idea of a First Cause, and the like. If
we say that every event must have a cause, it means
that every species of phenomenon belongs to a class
more comprehensive : that class to a still larger, and
so on. The "*summum genus*" of all (if any induction could reach it) would be nothing else than
an ultimate *physical* principle of the whole universe ;
but would still be so far from trenching upon the
idea of a supreme *moral* Cause, as to be, on the contrary, the very highest and crowning proof of the
influence of *mind*, in the evidence it would give of
the ultimate principle of *universal order*.

Conclusion.   The connexion and subordination of inductive
laws and generalisations is what we carefully distinguish as *physical causation*. But material *unity*,
system, and *order*, are the indications of *mind*; and
the connected series of *physical causation* is the manifestation of *moral causation*.

Thus, the truly inductive philosopher recognises
presiding Mind, the supreme *moral* Cause of all things,
everywhere revealed by the same outward manifestations of universal order and harmony ; everywhere
indicated by the same external attributes, symmetry,
uniformity, continuity ; and attended by the same

ministering agents, invariable laws, and physical
causes.*

---

* Hume's view of causation was censured by some of his opponents as
leaving the connexion of all events so loose as to open the door to the
supposition of causes sometimes failing to produce their effects, or effects
occurring without causes, or of all things being abandoned to chance or
destiny.

Though, as has been well observed by his biographer, Mr. Burton
(Life of Hume, i. 81.), such objections are of a vulgar class, and not such
as a philosopher would entertain, yet it may be worth noticing how completely the possibility of falling into such absurd misconceptions is
avoided by the view taken above.

NOTE TO PAGE 120.

It should be observed that the opinion quoted of Lord Kames, besides
the objection noticed in the text, involves also an instance of the confusion of the ideas of physical and moral causation here dwelt upon.

§ V. — FINAL CAUSES, AND NATURAL THEOLOGY.

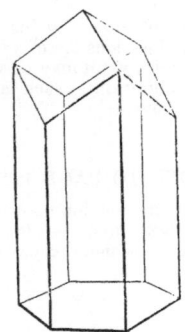

κατά τινα γεωμετρίκην μηχανῶνται προνοίαν.

PAPPUS.

"They work according to a kind of geometrical foresight."

Final causes.

THE theory of causation has been much mixed up with the argument evincing *design* in the arrangements of nature; and under the name of "Final Causes" that argument has been involved in no small confusion of ideas : and notwithstanding much which has been urged on the other side, still, with Dugald Stewart and others, I cannot but agree

The term ill chosen.

in thinking that the term "Final Cause" is most unhappily chosen to express the true meaning, and has tended to convey an idea not only too limited, but altogether confused and misleading.

Nor can I doubt that much of the obloquy which has been cast on natural theology generally by writers of a sceptical school, has arisen out of the narrow views thus implied, especially when the argument has been almost wholly restricted to physiology, and the very idea of intention represented as the essential characteristic of *organisation,* and this branch of science imagined to involve principles different from those prevailing in other branches ; all which might therefore naturally be imagined barren of such application.

It is, I conceive, solely from being understood in this narrow sense, that " final causes " are so vehemently assailed by Comte and writers of his school ; and it cannot be denied that among the advocates of natural theology there exists too common a disposition to *narrow* and *restrict* the application of the argument by *confining* the proofs of design to those instances *of adaptation of means to a perceptible end* of which we doubtless find such abundant instances throughout organised nature, instead of taking a more expanded view.

Objection from too narrow a view of final causes.

Means
adapted to
an end not
the highest
idea of de-
sign.

The instances in which we can trace a *use* and a *purpose* in nature, striking as they are, after all constitute but a very small and subordinate portion of the vast scheme of universal order and harmony of design which pervades and connects the whole. Throughout the immensely greater part of nature we can trace *symmetry* and *arrangement*, but not the *end for which* the adjustment is made. But this is in no way a less powerful proof of design and intelligence than the former. The most exact and recondite adaptation of means to accomplish an obvious end is *no more peculiarly* an evidence of design, than the universal arrangement according to determinate laws which pervades the depths of cosmical space,—

Symmetry
a proof of
design.

where we are least able to trace any end. Symmetry and beauty are *results of mind* of at least as high an order as mechanical efficiency. A mere numerical relation invariably preserved, but no further connected with any imaginable purpose, or a systematic arrangement of useless parts or abortive organs on a regular plan, are just as forcible indications of intelligence, as any results of immediate practical utility.

That the one class of results are more immediately

striking and obvious to minds of every capacity with little study or inquiry is true; and that the appreciation of the claims of symmetry even in unorganised structures, — the harmony of profoundly adjusted laws, — and the conspiring tendency of all these laws towards a grand pervading unity throughout the physical world,— is not so easily caught, and is perhaps fully appreciable only by the cultivated and philosophic mind, is equally true. But this is a difference merely in degree, and as applicable to different classes of minds. In a philosophic analysis of our convictions there can be no real difference in kind between the two classes of conclusions.

Proofs of design equally in symmetry and in an end answered.

In a strictly philosophical point of view the inference that everything "has a use" may certainly be regarded as a generalisation which carries with it a high degree of probability. We find that *many* things have a manifest use; but then we find innumerable others for which we can discover no use, and by which no visible end or purpose is answered, but it is not an unfair extension of the inference that, in these cases, some unknown end is answered; that, in fact, everything in natnre is adapted to other things and to the whole; though in

In most cases no purpose traceable.

by far the greater part of nature we fail to perceive *what* the particular relation or dependence may be. It may suffice to convince us of this, if we merely ask for what *purpose* is life itself conferred? or, to what *end* does the material universe altogether exist?

In some, purpose apparently defeated.

Again: the usual argument for design in organised structures is, that the various adjustments point to the designed end of life and enjoyment to which they are subservient; but it is an obvious objection that these ends in numberless cases are *not attained;* there is malformation and suffering, disorganisation and disease; and, finally, the whole design is always defeated and put an end to by death. It is hence manifest that to take a satisfactory view of the case, we must not rely on the *mere* consideration of an *end answered*, but must recur to a higher principle—that of symmetry, order, unity of plan, and composition of organised frames: and this too, as only one branch of the yet wider scheme of universal order.

But in certain cases an end answered when not obvious.

It is, however, fairly to be admitted, that many instances occur where we should least expect it of utility in natural arrangements. Thus it is argued by some that an *apparent waste* is not necessarily a *real* one ; for the sustenance of one single species or individual, a numberless combination of conditions must co-exist; if one condition were altered, all the rest would fail in their co-operation, and the individual, or even the whole species, would perish. Thus all, even the most apparently remote, arrangements of things which seem to have no relation to animal life are yet essential to it ; and thus the barren desert and the void ocean are not *wasted*, but essential parts in the economy of the minutest forms of animal life in the most distant hemisphere of our globe.

In the same point of view Œrsted has beautifully observed, " There is no inactive void in the remote distances between the planets. The space is filled by ether, and is penetrated by the attractive forces by which the whole universe is held together. The ether itself is an ocean whose waves form light, that great connecting link which conveys messages from globe to globe and from system to system." *

Yet the least consideration shows that we must

* Soul in Nature, p. 55.

L

not press such arguments beyond their due limits;
and still less make them the exclusive view of the
subject.

Views of
the ancients
on final
causes.

Though the ancients* reasoned justly and admirably on final causes in certain familiar instances,
and in the limited sense of the adaptation of means
to a known end, yet the state of their physical philosophy absolutely prohibited wider views of unity and
order.   Under a system which could not go beyond
the assignment of each class of phenomena to some
peculiar unknown efficient cause, unconnected with
others, no such generalisation as *unity of design*
could have been legitimately attained.

Final causes
misplaced
in philo-
sophy.

The remark of Bacon† that final causes are not in
themselves to be rejected, but have been *wrongly
placed* in philosophy, is one of more value than
seems generally understood.   It may be very true
that sometimes hints towards inductive investigation
have been obtained from the consideration of the
ends to be answered by certain observed conditions.
But it is in general a more safe and philosophical

---

* We cannot have a more striking instance than in the well-known
and justly admired passage in Xenophon's Memorabilia, i. 4.
. † De Augmentis, lib. iii. c. 4.

rule, that we may in all cases argue *from* physical inductions to final causes, but not *from* final causes to physical inductions.

The old and limited view of final causes will not meet the increasing demands of scientific enlightenment; it will not suffice now to argue solely on the adaptation of means to a known purpose, or a practical design evinced, and an obvious end answered. If we cannot discard the term, we must enlarge its meaning. We may speak of " design " with reference solely to " order " and " arrangement," without looking to the idea of *practical* utility. Such modes of expression are far preferable, as not leading the mind to any undue expectation of what it will not realise.

Improved views now required.

Thus in reference to physiology, the higher argument acquires an expansion in proportion to the progress of the science. We obtain more enlarged ideas of *design* as we advance from the more confined views of the older schools towards the wider principle of *symmetry and unity of composition.* So that " final causes," *properly understood,* so far from receding (as some pretend) before the advance of modern science in the wider and more philosophic

Extension of argument for design to unity of composition in physiology.

sense, eminently derive increasing evidence from its progress. The study of the higher principle of *symmetry* and unity of composition can in no way prejudice that of *adaptation ;* the latter being but a part of the same great argument. Nor is it just to accuse those of the modern school who are engaged, as their special and legitimate object, in investigating the former, of undervaluing the latter.

The cells of the honey-comb.

The celebrated case of the cells of bees deserves more particular consideration, inasmuch as it offers an instance in which the proof of *mind* is independent of the idea of mere utility. It is scarcely necessary to observe that the supposition adopted by some of a mere pressure upon a cylindrical cell producing the hexagonal form is wholly insufficient : the main point to be accounted for is the highly artificial mode of *termination* of the cell by three rhombs * inclined at the precise angle ($70° 31'$) which calculation requires for the minimum surface, which is also the acute angle of the rhomb. The argument points to a highly intellectual operation either performed by the bee, or implied in the arrangement of its organs, so

---

* See Vignette at the beginning.

as mechanically to effect it. On either alternative the proof of *mind* is independent of the consideration of a useful *end answered:* it depends on the conception and solution of what is to our intellects an abstract mathematical problem, by no means of an elementary or evident nature; and which is equally remarkable whether any *purpose* were *fulfilled* by its application, or not.

Paley expressly held that the mechanism of the heavens was a branch of science the least susceptible of this kind of application: according to the principle here advocated, it forms the highest and most satisfactory. *Argument from astronomy.*

But a more special argument has been raised on the ground that the planetary perturbations have been shown so to compensate each other, that no permanent derangement can arise; and Laplace pointed out that this stability of the planetary system is the necessary consequence of certain conditions, *not themselves necessary;* viz. the smallness of the inclinations and eccentricities, the motions all in the same direction, the comparatively vast mass of the sun, and the incommensurability of the periods. *Stability of the planetary system.*

Professor Playfair* justly enlarges on this as an argument for *design ;* but if the conditions thus assigned were necessary (i. e. *necessary consequences* of each other or of something else), he thinks we could not infer *design.* They, however, are not *necessary :* each *might be* otherwise, the rest remaining. Their existence then, he argues, not arising from *necessity,* nor from *mechanical causes,* nor from *chance,* must be from *design* and intelligence.

But I would ask, Suppose *they were necessary* consequences of each other or of some higher principle, or did arise from mechanical causes, would not that higher principle, or those causes, so arranged as to produce them, be an *equal* proof of *design,* or even a *higher?* So singularly deep-seated is the prejudice, that design can only be inferred when we cease to trace *laws,* or when conditions appear *arbitrary.*

Mechanical necessity.

The idea of "mechanical necessity" (derived probably from the school philosophy) as something distinct from the result of *systematic plan* in the order of the universe, has long continued to haunt the ideas of writers on the subject, and to be the source of many cavils.

---

* Playfair's Works, iv. 294. 318.

Thus in past times the Newtonian discoveries were accused by many of having an irreligious tendency in reducing everything to " mechanical necessity." And even so enlightened an advocate as Cotes*, instead of showing the fallacy involved in that very term, replies by contrasting "necessity" with "design," when it might have been pointed out that such *necessity of reason is the highest proof of design.* Other philosophers we find sometimes questioning whether certain results may have been brought about by the direct interposition of "the First Cause," or by some unknown "secondary cause," as if the two were *opposed* to each other; or, as if science could have any evidence of the first except through the channel of the second.

From the inductive philosophy we derive our belief in the harmony, order, and uniformity of natural causes, perpetually maintained in a universally connected chain of dependence. And hence it is, that we arrive at those sublime ideas of a presiding Intelligence of which *law* and *uniformity,*

Uniformity of natural causes.

---

* "Naturæ leges . . . in quibus multa sane sapientissimi *consilii*, nulla *necessitatis*, apparent vestgia."—Pref. to 2d edition of Principia, (p. xxix.).

universal mechanism once for all adjusted, are the proper *external manifestations.*

To the truly inductive philosopher, *fate* and *chance, necessity* and *accident,* are words without meaning. To him, the world is made up of recondite combinations of physical laws, and the existence and maintenance of those laws are the very indication of a Supreme Mind. But chance is irreconcilable with laws, fate with mind, regulated and fixed order with blind destiny, fortuitous accident, or arbitrary interruption.

Natural theology strengthened by the chain of causes.

All rational natural theology advances by tracing the immediate mechanical steps and particular processes in detail, and the physical causes in which the influences of the Great Moral Cause or Supreme Mind are manifested. The greater the number and extent of such secondary steps and intermediate processes through which we can trace it, the greater the complexity and wider the ramifications of the chain of causes, the more powerful and convincing the instruction they convey as to the existence and operation of the Divine wisdom and power.

Mistaken ideas.

Yet it is a common mode of illustration to speak of

the *chain* of secondary causes reaching up to the First Cause. Or, again, fears are entertained of tracing secondary causes too far, so as to intrench on the supremacy of the First Cause. But this is an erroneous analogy : the maker or designer of a chain is no more at one end of it than at the other. The length of the chain in no way alters our conviction of its skilful structure, except to enhance it. If the number of links were truly *infinite*, so much the more infinite the skill of its framer.

Mr. F. Newman * observes, I think most truly, that the *common* arguments from what are called "secondary causes" to the "First Cause" are unsatisfactory : and I would trace this to the confused sense in which those terms are commonly used, as already explained ; and which, I think, might be entirely removed by attention to the distinctions above laid down. While, on the other hand, I fully acknowledge that those arguments, when correctly understood, lead only to a *very limited* conclusion ; and one which falls infinitely *short* of those high moral and spiritual intuitions on which Mr. F. Newman

* Soul, p. 35.

grounds his religious system, yet in no way dis-
credits or supersedes them.

Again, by some well-meaning but confused reasoners, the argument is often put in reverse order;
and so stated as to appear as if the *assumption* of a
Supreme Mind or an " efficient and intelligent Cause "
were really the *basis* of our belief in the uniformity
of nature, instead of the *conclusion from it.* Yet if this
were so, what would it be but to render the whole
proof of a Deity *an argument in a circle ?* So, in like
manner, some would set out by insisting on the idea
of " a purpose answered " and " an intention " as an
essential antecedent part of our conception of an
organised being: and then, from the study of organised beings, would *deduce* the *conclusion* of design
and intention !

Coleridge observes, " Assume the existence of
God, and then the harmony and fitness of the physical creation may be shown to correspond with, and
support, such an assumption : but to set about
proving the existence of a God by such means, is a
mere circle,—a delusion ! " * Now I would observe

---

* Table Talk, p. 307.

that for the theological idea of God, the natural argument is no doubt *insufficient,* but still it is no argument *in a circle;* it is strictly logical as far as it goes, though that is but to a very limited extent.

Again, the same author asks, " How did the Atheist get his idea of that God whom he denies?"* The answer is unhappily obvious, that he usually takes it up from the narrow and unworthy representations of dogmatic systems or puerile recollections, instead of the inferences suggested by a sound philosophy, which would dissipate his objections.

Among some writers of an eminently religious spirit at the present day, we cannot but notice the unhappy influence of that confusion of ideas on the subject of *causation,* as well as the want of due appreciation of the grounds and nature of physical philosophy in reference to the inferences of natural theology, which it has been the object of the foregoing remarks to obviate. Mistakes arising from ideas of causation.

Thus Sterling† observes, " Physical results prove nothing but a physical cause." Again, " It is

---

* Table Talk, p. 307. † Essays, ii. 121, 122.

thoughtless to say that, because all things we know have each their cause, therefore the whole must have one cause." . . . . . " Every phenomenon *within nature* has a cause ; but this does not entitle us to go beyond and look at nature from without, and say this, too, must have a cause." *

The real ground he maintains is very different, and is suggested by the question, " Why is the view of the universe so weary, fearful, and unsatisfactory ? The sense that we need a God is an infallible indication that there is one." . . . . . &c.*

After what has been before observed, it is hardly necessary to observe how completely all perplexity would be removed by better views of physical philo-

---

* It has been represented to me by a friend of the late Mr. Sterling, that, in what I have here said of him, I appear to misconceive his meaning, and that in fact what he says really seems to agree closely with what I have myself afterwards urged, Essay iii. § 3., " Legitimate science," &c.

But perhaps it will be observed that what I have there said refers rather to the *past*, and the passage now in question to the *present*. The main question is *in what sense* Mr. Sterling here uses the term " cause." This is, I think, the source of the difficulty. If in the sentence " Every phenomenon," &c., in both places he means " physical cause," I should entirely agree with him. But the tenor of the whole seems to me to indicate that he is speaking of " efficient causes ; " in which case I should differ. But I still think there is some of the very common confusion between *physical* and *moral* causation.

" Everything within nature has a [physical] cause." This, I think, *does* justify us in concluding that nature, as a whole, has a *moral* cause ; — it is the very evidence of it. The notion of a *moral* cause to which I refer is nothing else than what arises necessarily out of the conception of the vast assemblage and orderly combination of *physical* causes. As to any idea of personality, power, or moral attributes — all these I entirely agree must be derived from quite other sources, as they are conceptions of a totally different order.

sophy. While with sincere admiration for the author's literary attainments, poetic imagination, and high devotional feeling, I cannot but think the unhappy view of the universe as " weary, fearful, and unsatisfactory," would have been banished at once by the juster contemplations of an enlarged inductive philosophy, investing the whole with the cheering light of universal beauty, order, and harmony.

If, indeed, the author meant simply to transfer the belief in a Deity altogether from the domain of *reason,* to place it in that of *spirit;* to ground it on the sole consciousness of internal emotion, or the intuitive impressions of individual experience, this would be a view to which the philosophical argument *offers no disparagement,* though it does not reach up to it.

Higher ideas of God from other sources.

A Personal God,—a moral Governor of the world, — the Divine Will and Power originating material things, and calling forth intellectual and spiritual life, are doctrines *not of science, but of faith,* and repose on the same ground as all other religious doctrines. As to the nature of those grounds, they will necessarily be different in the case of different individual minds. But in point of fact, it is, I imagine,

the case that by far the larger majority derive such
conceptions from the language of the Bible, instilled
into their ears and memory from the earliest childhood; though doubtless there are many who adopt
them from higher spiritual impressions and internal
feelings and convictions;—but in either case from
sources wholly distinct from the teaching of science.

Erroneous
notions of
the limits
of nature.

Some, however, would assert, that after all physical
explanations, there remains the same ultimate incomprehensibility in natural causes; and that even in
nature we find ourselves surrounded by wonders and
miracles : ideas which only evince a total absence
of distinct philosophical thought, and confound the
limits of *nature* with the limits of our PRESENT
knowledge of it — *unexplained* phenomena with *violations* of physical order.

They are fond of speaking of the *limits of nature,*
of a region of inscrutable *mystery* by which the
frontiers of science are on all sides surrounded, im-
penetrable to our faculties, and forbidding advance.

No hin-
drance to
advance.

If by mystery they mean something into which
we neither can, nor ought, to inquire, then, in accordance with what was before observed,* in science there

---

* See above, p. 112.

are no mysteries, no inductive inquiry can ever bring us to such a termination.

If we limit the term " nature " to that portion of the universal fabric whose laws and mechanical causes are more or less perfectly known to us, the distinction is then merely incidental and fluctuating : it is purely *relative* to ourselves and the temporary extent of our knowledge, and presents no really essential difference, being dependent merely on the extent to which the boundaries of knowledge are pushed forward at any particular epoch.

But " nature," in its wider sense, implies a whole, all of whose parts are united by a community of character ; and no one portion of it, whether known or unknown to us, can really be beyond those ordinances of recondite arrangement, a small portion of which is manifested to us.

Nature means universal order.

To assert an arbitrary condition of things whenever our inductions fail, is to place such cases beyond the boundaries of design : to suppose a region of mysterious confusion beyond all law and order, is to discredit the universal influence of Supreme Mind.

Some persons look to a supposed limit of all physical knowledge, a supposed boundary of the dominion of physical causes; there to enter on a new region, still within the domain of reason; and then to discover the evidences of the presence and majesty of the Deity. They imagine it is only when they arrive at the termination of natural order, that they can properly say " Deus intersit; " and yet that this is still part of the province of science.

But reasoners of this class are liable to perpetual disappointment, whenever, as daily happens, the ceaseless progress of discovery pushes forward the apparent boundary of any such limitation temporarily placed on our knowledge of nature, by disclosing a new region of facts, converting what was before obscurity and mystery into clear light, and opening a wider horizon to our contemplation.

The argument of Natural Theology, instead of being supported or enlarged by such mistaken yet prevalent imaginations, is on the contrary exposed to continual disparagement, failure, and defeat, so long as its advocates persist in relying on such false supports.

While the real argument is continually deriving fresh accessions of strength from every higher advance in generalisation which is effected, and continually raising the ideas of those who accept its conclusions towards

higher and more worthy conceptions of the Supreme
Moral Cause.

In accordance with the narrow and unworthy
notions formerly prevalent on these subjects (perhaps inseparable from an earlier stage of science), it
would seem to have been held, that the appearances
of the physical world, so far as they were reducible
to regular laws, were to be regarded as what was
termed " Nature." When we reached the boundary
of the province thus subject to reason (as we soon
must do in any direction), and when phenomena
seemed in any instance not so reducible to laws,
then we arrived at the limits of " nature," and were
reluctantly compelled to resort to a Deity, a θεος απο
μηχανης,—a Supreme Being admitted on compulsion,
when the order of things could no further be traced
without Him. Then, and not till then, we might
exclaim with the poet, " Ergo perfugium." *

Limited notion of " nature," as opposed to " Deity."

Thus, to take an instance, minds incapable of appreciating Newton's own sublime inference from the
uniformity and order of the system which he had so
marvellously and happily disclosed, have dwelt upon

Erroneous argument from planetary perturbations.

---

* Lucret. v. 1185.

M

his single expression (when in ignorance of the extent
and fertility of his own principle in leading to the
great law of stability), that at length the increase of
planetary perturbations would require a special intervention for restoring the equilibrium. This was ap-
plauded as the only satisfactory acknowledgment of
a Supreme Power. We merely ask, *If this be the
true argument, what now becomes of the conclusion ?*

Certain phenomena in geology.
     Just in the same way we hear (for example) re-
ligious writers at the present day arguing on certain
obscure and unexplained phenomena of geology.
They find indications of what may seem *apparently*
abrupt changes in the orders of organised beings in
past times: and because no established law or physical theory will immediately apply to assign an
adequate cause (supposing the fact to be so), these
changes are triumphantly adduced as the special
footsteps of the Creator (as if the whole of geology
presented anything else); so that when future and
enlarged discovery shall disclose the connexion and
explanation of these appearances by regular laws,
their argument for a Deity will fall to the ground!

In physics.
     According to this mode of representation, "nature"
was the *rule*, "Deity" the *exception*. The belief in

nature was the doctrine of reason and knowledge; the acknowledgment of a God was only the confession of ignorance. So long as we could trace physical laws, nature was our acknowledged and legitimate guide; when we could attain nothing better, *we were to rest satisfied with a God!* Even learned writers on natural theology have thought it pious to argue in this way. To take a single example : The apparent anomaly that water arrives at its maximum density before freezing, occasions its *freezing first at the surface*, and other results connected with important points in the economy of the globe and the good of its inhabitants : and this argument for design is sometimes represented as if it acquired a *peculiar force* from the circumstance of the fact being an *anomaly*, and inexplicable by our theories. And *on this ground* it is particularly held up to popular acceptance as an instance of special intervention, for the benefit of man, traceable to no physical cause. But when the apparent exception shall come to be reduced to its proper place as a part of some more comprehensive law (as it assuredly will)*,

---

* See above, Essay II.

all the peculiarity and mystery of the case will be at an end, *and with it will fall the* theological argument and the popular faith, propped up on so false a support.

Yet in spite of the better knowledge which ought to prevail, we often hear, for example, any sudden and marvellous infliction of disease or famine, pestilence or blight, which (it is added with a sort of triumph) " baffle the boasted powers of science to explain," held forth as signal instances of direct interposition.

To resort to such representations, however it may serve a temporary purpose, or exert an influence on the multitude, is the resource of ignorance, the encouragement of superstition, and eventually the unfailing parent of a sceptical and irreligious reaction; and if the faith of the many be propped up by such false supports, it must fail altogether as soon as increasing knowledge clears them away.*

To speak of apparent anomalies and interruptions

---

* In relation to this subject, I cannot refrain from quoting a single sentence from a discourse of a very opposite tendency to such as I have just alluded to, and of a kind which it is to be wished were more common : —

" God punishes us . . . . not by His caprice, but by His laws. He does not break His laws to harm us; the laws themselves harm us when we break them and get in their way "

" Who Causes Pestilence ? " Four Sermons by the Rev. C. Kingsley, Rector of Eversley, p. 14. London, 1854.

as *special* indications of the Deity, is altogether a mistake. In truth, so far as the *anomalous* character of any phenomenon can affect the inference of presiding Intelligence at all, it would rather tend to *diminish* and detract from that evidence. But, on the other hand, precisely in proportion as the apparent exception might be explained, and made to vindicate its position in a more comprehensive system of order, so would the evidence be increased and elevated.

In the present state of knowledge, law and order, physical causation and uniformity of action, are the elevated manifestations of Divinity, creation and providence. Interruptions of such order (if for a moment they could be admitted as such) could only produce a sort of temporary concealment of such manifestations, and involve the beautiful light shed over the natural world in a passing cloud. We do not indeed *doubt* that the sun exists behind the cloud, but we certainly do not see it; still less can we call the obscuration a special *proof* of its presence. The main point in the system of order and law is its absolute *universality.* Exceptions, if real, must *pro tanto* imply a deficiency in the chain of connexion,

M 3

and might, to a sceptical disposition, offer a ground of doubt.

But so overwhelming is the mass and body of proof, that no philosophic mind would allow such exceptions for a moment to weigh against it; they would be as dust in the balance. A supreme moral cause manifested through law, order, and physical causes, is the confession of science: conflicting operations, arbitrary interruptions, abrupt discontinuities, are the idols of ignorance, and, if they really prevailed, would so far be to the philosopher only the exponents of chaos and atheism; the obscuration (as far as they extend) of the sensible manifestation of the Supreme Intelligence.*

---

* The question of apparent interruptions has been much discussed by Œrsted, in his work already referred to. (See Soul in Nature, pp. 59. 173. 178.) Some, he says, imagine accidental causes of derangement in nature which may render arbitrary intervention necessary for their readjustment, as was once supposed with regard to the planetary perturbations; but, as knowledge has progressed, it has been more clearly, seen that the error lies in supposing such deranging causes accidental; they are all results of the same general laws, modified by particular conditions. He puts the parallel of human contrivances, in which, in proportion to the skill and intelligence employed, such derangements are foreseen and provided for, as e. g. the compensation for expansion in chronometers; and even in moral agency, as in state institutions providing remedies for the lawlessness of criminals. (173.) Infinitely more then, he argues, must we expect such provisions in *Infinite Wisdom* which is sufficient to guide everything without requiring alteration. (178.) Apparently inexplicable events are so only to our present ignorance. (59.) Real interruptions would suppose deficiency of reason in God.

On this point the author has forcibly remarked, "When our opponents triumphantly bring forward inexplicable events, we can reply to them,

That idea was hidden in former ages, ignorant of the true inductive philosophy, — in proportion as men speculated on false principles, and imagined (as was very natural) a plurality of supernatural powers, and unconnected or opposing influences, whose conflicting operations were evinced in the mysterious irregularities and capricious course of natural phenomena. In earlier stages, even of inductive inquiry, though its higher principles were in some measure recognised, it was yet supposed that limits existed to their dominion. And even at the present day we cannot say that such a notion has been generally or absolutely exploded.

False views from ignorance of inductive principle.

Among philosophers, though the idea of a limitation has been slowly dispelled, yet still many are unable fully to embrace the universality of law and order, or to attach to it the high importance which it rightly claims.

Narrow views of nature to be enlarged.

In common with us you cannot understand these events, but you fancy you understand them ; you believe that you are initiated into God's decrees, and speak accordingly: we know that we do not understand them, and openly declare it. They may perhaps assert that they are guided by religion; that they judge by the will of God revealed to them by religion: but only let them show us a single instance of an event where it can be applied without the addition of some of their own wisdom." — Ib. 178.

" We often hear it said that some things would be inexplicable if we did not believe in higher arbitrary arrangements; but that anything is inexplicable without a certain presupposition is generally a very weak proof of its being really so."— Ib. 180.

In a word, according to the narrow, but prevalent idea, the great universe is nothing but an immensity of arbitrary and inscrutable darkness and mystery, to which the philosopher's inductive world — a limited portion of matter, chained down by mechanical causes—forms an insignificant exception.

But the more worthy conception looks to a boundless cosmos, a universe of order, a grand scheme of eternal laws adjusted by Supreme Reason, of which our limited inductive knowledge opens a partial glimpse, yet calculated to convey a faint impression of that immensity of Intelligence which pervades, animates, and rules not only our sphere, but all beyond it.

Improvements in science advance natural theology.

Improved views, increased and accumulating evidence of the harmony pervading the material world, are attained in proportion to the advance of sound inductive science. The more close adherence to the spirit of philosophical analogy leads to a more commanding sense of the uniformity of nature, and the true idea of causation. As the generalisations of physical science become more comprehensive, we acquire juster notions of the stupendous aggregate of physical causes, of the inconceivable vastness and complexity of that universal mechanism, some small portion of which we are enabled to understand; and whose recondite and perfect adjustment, however imperfectly perceived, is the true ground and evidence of our conceptions, partial and limited as they must be, of the Infinite Source of all things.

It is not a mere desultory and fragmentary knowledge of detached facts and portions of science, to however great an extent it be carried, which can suffice to lead us to a correct perception of those truths. It can only be by a thorough insight into the interior principles of the inductive philosophy, and an imbibing of its real spirit, that we can attain an adequate perception and sense of the real unity of nature which forms the basis and substance of those more sublime inferences.

In the confined and literal notions, often ignorantly entertained, of the sciences of observation, our conclusions might be supposed restricted to the field of mere sensible experience; and in this sense we should fall short of any worthy apprehension of the Supreme Intelligence. But the truly inductive philosopher extends his contemplation to intellectual conceptions of a higher class, pointing

Higher views of science important.

Leading to higher views of natural theology.

to order and uniformity as constant and universal as
the extent of nature itself in space and in time; and
in the same proportion he recognises harmony and
arrangement invested with the attributes of universality and eternity, and thus derives his loftier ideas
of the Divine perfections.

The real nature and bearing of the evidence of
natural theology as founded on *universal order,* has
in fact come to be better understood only in an age
of advanced philosophic cultivation: it tends to
become continually more perfect with increasing
knowledge; and its full force is hardly yet commonly apprehended even among men of science.

In ignorant ages all phenomena viewed as supernatural.

The stupendous phenomena of nature are indeed
the manifestations of the Supreme Power as well
to an ignorant, as to a cultivated age and people,
though the impression is produced in a very different way, and excites a very different tone of feeling. In an age when these phenomena have not
been reduced to laws, or traced to causes, they are
all ascribed to arbitrary influences. When the uniformity of nature was unknown, violations of it
offered no contradiction. When everything was
supernatural, no discrimination of evidence was pos-

sible. In those ages all phenomena, whether in the inorganic world, or in varied influences on the human constitution, were unavoidably regarded as direct interventions and acts of the Deity, and were truly described as such in the writings of those periods.

In the ruder stages of man's progress, religious impressions are more peculiarly produced through the medium of the feelings of awe and astonishment, which are called forth by the occurrence of the more rare, extraordinary, and fearful phenomena, prodigies and marvels. These joined to the more tangible influence of events on their own fortunes and enjoyments, believed to be retributive judgments and providential deliverances, when favourable to themselves and destructive to their enemies, are the only appeal to which men were then accessible. *Natural theology of ruder ages.*

But with the enlightenment of physical discovery, a more definite natural theology presents us with conclusions which, though resting on an unassailable basis, are restricted in their character and extent. *More enlightened views.*

Natural theology, as based on physical science, confessedly leads us only to a very limited conception of the Divine perfections; it traces beneficent arrange- *Limits of natural theology.*

ments yet mixed with a large proportion of evil; it recognises omnipotence in the constitution of the immense connected machinery of the universe, and the perpetual maintenance of determinate laws, rather than in their interruption. At the very utmost it points to providential government in the preservation of an unbroken system of pre-ordained causes for the general good, rather than its suspension for the benefit of individual parts, and to influence on mind rather than disarrangement of matter.

Spiritual views from other sources.

If the human mind or human desires require fuller manifestations, or aspire to a higher sense of the Divinity, it must be from *other* and *more spiritual sources* that such wants can be satisfied,— a philosophic natural theology, while it cannot furnish such satisfaction, yet at least puts no hindrance in the way of its attainment from other and more appropriate teaching: but rather prepares the way for it by clearing away unworthy notions which obstruct its path.

Argument from design to a designer.

But the great argument which we have been considering, it is said, is not one *merely* of *design*, but must rise *from design to a designer.* And here it is that some objections have arisen.

On the one hand it is alleged that the argument
is insufficient; and, on the other, that it proves too
much, and tends to identify nature with the Deity.
But both objections seem to me equally traceable to
the same primary confusion of ideas as to the real
nature of the inductive inferences, and of the obvious
distinction between moral and physical causation.
This confusion of ideas pervades the remarks of many
otherwise able writers.   Thus Coleridge observes,—

"All the so-called demonstrations of a God either
prove too little, as that from the order and apparent
purpose in nature, or too much, namely, that the
world is itself God; or they clandestinely involve
the conclusion in the premises . . . as in the postulate of a First Cause." *

Natural theology confessedly " proves too little,"
because it cannot rise to the metaphysical idea or
scriptural representation of God.   These stand on
quite distinct authority.   But " the postulate of a
First Cause " is a notion wholly arising from the
confusion of ideas just referred to.

The common objection to the argument from Objection to
this argu-
ment.

* Aids to Reflection, vol. i. p. 139.

design to *a designer*, appears to be of this kind.
It is alleged that, to take Paley's well-known instance of the watch, we make our inference directly
of a *watchmaker* from obvious comparison with
known human works.   Even when a person should
for the first time witness some work far transcending
his own power or knowledge, or anything previously
heard of, still he would perceive the analogy with
the more ordinary productions of human skill, differing only in *degree*, and would infer a contriver
and an artist of faculties far higher, but still similar to his own.   But the works of nature, it is
said, differ from these *in kind;* they are unlike any
of our works, and suggest no such analogy of an
artificer resembling a human artificer, or differing
merely in the extent and degree of his skill.

Apparent
want of
analogy in
the two
cases.
In those cases most nearly approaching the nature
of human works, such as the varied and endless
changes in matter going on in the *laboratory of
nature,* the results, even when most analogous to
those obtained in human laboratories, yet present no
marks of the process or of the means employed, by
which to recognise the analogous workman; and in
all the grander productions, the incessant evolutions

of vegetable and animal life, which no human laboratory can produce, — in the structure of earth and ocean, or the infinite expanse of the heavens and their transcendent mechanism, still further must we be from finding any analogy to the works of man, or by consequence any analogy to a personal individual artificer.

But the more just view of the case is that which arises from the consideration that the real evidence is that of *mind* and *intelligence :* for here we have a proper and strict analogy. *Mind* directing the operations of the laboratory or the workshop, is no part of the *visible apparatus,* nor are its operations seen in *themselves* — they are visible only in their *effects ;* — and from effects, however dissimilar in magnitude or in kind, yet agreeing in the one grand condition of *order, adjustment,* profound and recondite connexion and dependence, there is the same evidence and outward manifestation of IN-VISIBLE INTELLIGENCE, as vast and illimitable as the universe throughout which those manifestations are seen.

Answered by the real analogy of mind.

It is by *analogy* with the exercise of intellect, and the volition, or power of moral causation, of

Analogy of moral causation and mind.

which we are conscious within ourselves, that we
speak of the *Supreme Mind* and *Moral Cause* of the
universe, of whose operation, order arrangement
and adaptation, are the external manifestations.
Order implies what by *analogy* we call *intelligence;*
subserviency to an observed end implies intelligence
foreseeing, which, by analogy, we call *design.**

Invariable
laws and
causes no
argument
for Panthe-
ism.

Again: nothing but the common confused and
mistaken notions as to laws and causes, could give
any colour to the assertion that " the argument proves
too much," that physical speculations tend to substitute general physical laws in the place of the Deity;
and that scientific statements of the conclusions
of Natural Theology are nothing but ill-disguised
Pantheism.

The utter futility of such inferences is at once
seen, when the smallest attention is given to the
plain distinctions above laid down between " moral "
and " physical " causation: and to the proper force
of the conclusions from natural science establishing
the former by means of the latter.

---

* On the *analogical* nature of all our modes of speech respecting the
Divine Being and attributes, the reader should especially consult the
luminous and philosophical view given by Abp. King in his " Discourse
on Predestination," 1709, § 8. Reprinted 1821, with Notes by Archbishop Whately.

This distinction obviously points to the *very reverse* of the assertion that physical action is identical with its moral cause; the essential difference and contrast between them is the very point which the whole argument upholds and enforces.

Of all forms of philosophical mysticism, the idea *Pantheism.* of Pantheism seems to me one of the most extravagant. Ever-present *mind* is a direct inference from the universal order of nature, or rather only another mode of expressing it. But of the *mode* of existence of that mind we can infer nothing.

To assert, then, that this universally manifested mind is *coexistent*, or even to be *identified*, with *matter*, is at best a mere gratuitous hypothesis, and as such wholly unphilosophical in itself, and leading to many preposterous consequences. But if further supposed to apply in any higher sense as to an object of worship, trust, love, obedience, or the like (as is implied in the term Pan-*theism*), it appears to involve moral contradictions of the most startling kind.

There are, however, many who, though rejecting Pantheism as *untrue*, do not conceive it *absurd* or *contradictory*. Much, however, will, in all such cases,

N

depend on the precise *sense* in which it is maintained. With some it seems to have been upheld on a fanciful analogy with the conception of the human frame animated by an indwelling spirit; as if in a somewhat similar manner the supreme mind might animate nature. Without disputing this in a certain sense, the cases surely cannot be considered at all parallel: we do not infer the existence of the human mind from the arrangement and adaptation of the bodily organs: nor is it the moral cause of their organisation.

If Pantheism were asserted merely in the sense of a kind of vital or animating principle pervading the material world, I would admit that such an idea involves *no absurdity,* or *contradiction,* but still I should regard it as visionary and unphilosophical. I could but class it with the "vital forces" which Kepler fancied necessary for keeping up the motions of the planets,—with " the plastic powers of nature," " her abhorrence of a vacuum," and the like chimæras. But it is when men elevate such a supposed animating principle into a *Deity,* a being of supreme wisdom, power, beneficence, and goodness, yet residing in every atom of matter, and *participating directly* in every form and case of material action, that the contradiction arises.

The whole tenor of the preceding argument is Conclusion. directed to show that the inference and assertion of a *Supreme Moral Cause,* distinct from and above nature, results immediately from the recognition of the eternal and universal maintenance of the order of *physical causes,* which are its essential *external manifestations.*

Of the *mode of action or operation* by which the Supreme Moral Cause influences the universal order of physical causes, *we confess our utter ignorance.* But the *evidence* of such operation, where nature exists, can never be lost or interrupted. And in proportion as our more extended researches exhibit these indications more fully and more gloriously displayed, we cannot but believe that our contemplations are more nearly and truly approaching their SOURCE.

ESSAY II.

ON THE
UNITY OR PLURALITY OF WORLDS.

§ I.—THE ARGUMENT CONSIDERED IN A PHYSICAL AND PHILOSOPHICAL POINT OF VIEW.

LORD ROSSE'S TELESCOPE.

AMONG the endless topics of human inquiry and controversy, we cannot but observe that those often become the most popular which might have been antecedently pronounced the most unlikely to attract notice : and subjects the most remote from those which ordinarily engross the attention, involve the interests or excite the passions of mankind, and even those of the most imaginary and conjectural character, have often called forth the most earnest dispute. *Introductory remarks. Propensity to discussion of remote and imaginary topics.*

Often, too, it happens that a particular opinion has for a long time been conventionally adopted, or *Love of novelty.*

acquiesced in—perhaps with little inquiry or thought among the many—until on a sudden some writer, more bold or more ambitious than his predecessors, discovers a dubious point on which at least a plausible argument may be raised, calling in question the received belief in which the public mind has hitherto reposed.   And then, however abstract, or however trivial the subject may be, prepossessions are aroused, and the question is immediately raised into serious importance, and a controversy stirred up whose vehemence is often just in the inverse ratio of the real value or clear evidence of the point at issue.

Interest felt in conjectural speculation.

Men who take comparatively little interest in the tangible details of real attainable science, feel stimu˘ lated by the desire to penetrate those recesses where all is obscure, and *certain* knowledge unattainable. There is indeed a line of demarcation, nowhere more clearly drawn than in the positive mathematico-physical sciences, between the known and the unknown; yet on the frontier there lies a region on which sufficient light is shed to permit our curiosity safely to indulge in short incursions, while there is darkness beyond which we may people with unsubstantial forms and shadows to the full satisfaction of our wildest imaginations.   And this is pre-eminently ₜhe case in the vast expanse opened to our view by astronomy; and when readers and inquirers are tired with the dry discussions of periods and distances, and with calculations of masses and eccentricities, they naturally fly for relief to the more grateful occupation of guessing at the probable near aspect of the surfaces of the heavenly bodies —imagining them worlds like our own, and fancying the possible nature of the creatures who may be their inhabitants.   And as such subjects are of course open to unlimited conjecture of every kind, so are they liable to become the battle-field of interminable dispute : interminable, perhaps, some may say, in proportion as the hosts of objections raised on one side, and of replies on the other, are all equally unsubstantial ; and like the aërial combatants of the poet, of impassive and indestructible nature, reuniting as soon as cloven asunder, so that, after all is over, either side may with equal confidence claim the victory, and be ready to begin the fight anew.

Yet, to the question of the existence of inhabitants in other worlds (so impossible really to

Popular belief in a plurality of

answer decisively) there seems to have prevailed a general disposition to give in some sort an affirmative reply; not, perhaps, founded always on very scientific grounds; but undoubtedly the common expression of popular books has been favourable to the idea, and perhaps most educated persons would hardly have hesitated to admit its general probability.

In fact, no sooner was the true planetary system generally received, than it became an obvious topic with writers who were engaged in recommending it to public notice, to invest the dry details of planetary astronomy with the interest excited by speculation on the possibility of intelligent beings inhabiting those distant worlds. And when men had become reconciled to the paradox of *antipodes*, it was not a much greater difficulty to concede *Lunarians, Jovians, Saturnians,* and the rest. And thus the general idea of inhabited worlds, under various aspects and with various objects, has been alike upheld by philosophers and poets — by divines and popular essayists. Hinted at even by Newton and Huyghens — recommended to popular acceptance by the elegant discussions of Fontenelle — reasoned upon theologically by Derham and Bent-

Inhabited worlds.

Views of the subject in former times.

ley, it passed into a matter of popular credence; and was often appealed to even in later times, both as a fair philosophic hypothesis and as a worthy religious contemplation by Herschel and Paley — by Lardner and Chalmers, and by Dr. Whewell in his Bridgewater Treatise.

At the present day, it hardly needs to be re- *Discussion at the present day.* marked, a contest on this subject has been keenly carried on between two highly talented disputants, whose publications* have called forth an unusual amount of popular attention; and the question of the *plurality of inhabited worlds,* which had altogether slumbered in public interest, as a controverted point, since the days of Fontenelle, has once more started into life, and occupied the public mind; and after the public scepticism had been at length satisfied as to the *actual rotation* of the earth by the pendulum experiment of M. Foucault, this discussion took its place as a fashionable topic, and commended to the public favour the Copernican hypothesis; on the one

---

* 1. "The Plurality of Worlds," an Essay: 1853.　2. "A Dialogue on the Plurality of Worlds," being a Supplement to the Essay, &c. 3. "More Worlds than One; the Creed of the Philosopher, and the Hope of the Christian," by Sir David Brewster, K. H., D.C.L., F.R.S., V.P.R.S. Edin., and Associate of the Institute of France, 1854.

side, as perfectly *safe*, when carefully divested of
the dangerous adjunct of imagining inhabitants in
the planets rivalling man in dignity and spiritual
privileges; on the other, as eminently orthodox, if
the planets be believed to be tenanted by such inhabitants, whose existence is even demanded by
religious considerations, and made at once a leading
point in " the creed of the philosopher and the hope
of the Christian."

In the instance of the present controversy, whatever opinion may be formed as to the issue or the
merits of the question, there can be but one as to
the ability with which each disputant has conducted
his argument; and especially as to the ingenuity
which the anonymous writer has evinced in maintaining what must seem in many respects somewhat
paradoxical theories, the more freely thrown out
under his anonymous disguise, doubtless assumed
for this purpose.

Various
ideas
broached of
the inhabit-
ants of
planets.

The literature of such a question is not without
its curiosities; and both the authors referred to have
contributed some account of the varied opinions
which have been broached on the subject. The
author of " The Plurality " displays his stores of

erudition on this head in furnishing several singular exemplifications.

Some of the ancient Stoics, as we learn from the ridicule cast on them by Lactantius, ascribed inhabitants to the moon. Plutarch, in his curious dialogue "on the face which appears in the moon's orb,"* gives arguments for and against the moon being inhabited. Lucian indulges in the same fancies; but probably neither with any settled conviction of their truth. Nicolas of Cus, who asserted the Heliocentric system before Copernicus, discusses in some measure the nature of the solar and lunar beings, and makes the former more intellectual, more clear and illuminated, than those of the moon, who are " magis lunatici," as in the earth they are "magis materiales et grossi."

These were followed by the strange assertions of Giordano Bruno, who declared not only for a plurality of worlds, but that the earth is inhabited in its interior as well as its exterior; and Wilkins, who laboured to prove not only that the moon is inhabited, but that we need not despair of being able

---

* Supplement, p. 36.

to visit the inhabitants.   Kepler expressly argued in favour of Lunarians : " Consentaneum est esse in lunâ viventes creaturas."

Even down to later times, imagination has revelled in devising the kind of existences which may inhabit the planets. Fontenelle assigned corporeal beings very like ourselves to the nearer planets; creatures of extreme vivacity to Mercury; of voluptuous and ardent natures to Venus ; of more robust and manlike character to Mars ; while to Jupiter and Saturn beings of dull and torpid constitution were given. Sir H. Davy, in his vision, saw creatures of the most marvellous structure, with membranous bodies and strangely convoluted elephantine probosces as organs of sense and intelligence, gifted with far higher intellectual capacities than the men of this earth, inhabiting Saturn; to which we may add, that a very ingenious and scientific poet has recently depicted beings in the moon with an *internal* body and an *external* soul.*

It is not, however, to any such familiar acquaintance with other worlds that our immediate inquiry

* "Love in the Moon," a poem, with Remarks on that Luminary, by Patrick Scott. London, 1853. See p. 24.

is to be understood to aspire. The actual point in question is not what is the most plausible image we can conjure up of the nature and appearance of lunarians or planetarians, but what is the most proper and philosophical view we can take of the general question, *Are other worlds besides our own* PRO-BABLY *the seats of intellectual, moral, and spiritual life?* Is it probable from concurrent circumstances that our globe is so far a peculiarly conditioned portion of the whole creation as to be the only one privileged in this respect? or are not others, or all others perhaps, equally, or even more, elevated in their destination as seats of life?

The question is one which most persons at all Statement of the ques tion. given to contemplate the phenomena of the heavens have been always prone to ask, though perhaps with little serious conviction of the possibility of answering it with any reasonable approach to certainty. But it takes a more precise form, and the real nature of the inquiry is more distinctly indicated, the more we consider the actual conditions of the sidereal world. The question, probably, first arises with respect to those bodies nearest to us, and which most resemble our earth. From the obvious General analogies

general resemblances between our globe and those
others of the solar system which might be regarded
as members of the same family in respect to form,
motion, subjection to the laws of gravitation, diffusion of light and heat, in some instances the
presence of moons or rings affording auxiliary
illumination, in others of atmospheres, clouds, and
therefore water, of mountains and valleys, or even
of supposed continents and oceans, there might
seem to be an easy transition to the belief in intelligent inhabitants, bearing more or less resemblance
to ourselves.

The inquiry perhaps is first made, — is our com-
panion the moon inhabited ? It then extends to the
other planets, and as it extends, it may seem surrounded with more difficulty; are we to include the
sun ? or is not his nature so different as to render
such an inquiry unreasonable ? Are comets likely
to be inhabited ? still more the fixed stars ? but
they probably are suns; they may be the centres
of planetary systems — of worlds; why not of
animal life, or even of intellectual and moral life ?

But there are yet more distant bodies, the systems
of nebulæ and clusters, to which the same questions

may apply; and thus the overwhelming magnitude of the inquiry makes us the more alive to its difficulties, and we feel ourselves lost in the vastness of the conceptions it inspires.

At the present day, from more accurate investigations, aided by the recent improvements in telescopes, the actual structure and physical conditions of the planetary bodies of our world, as well as in some degree those of the more immensely distant and vast sidereal systems, have been better known to astronomers, and correct ideas respecting them more familiarly diffused in popular information. Hence it becomes a point of inquiry whether these accessions to our knowledge have been such as in any way to affect the previously received notions as to the existence of organised life in the heavenly bodies.

Influence of modern discoveries on the question.

To follow up this inquiry is the professed object in a large portion of the discussion now brought before the public; and it is apparent, that while the modern discoveries generally have confirmed and extended the analogies of planetary and stellar systems, they have also disclosed many particulars which may require us to modify our notions in detail as to the conditions of their existence; while geological research has not been without its bearing on the question of their structure, nor the various cosmical and cosmogonical theories altogether uninfluenced by the latest discoveries of nebular astronomy.

It was thus but fair and reasonable that the question should at the present day undergo a renewed discussion; and whatever opinion may be formed as to the precise result to which the present controversy may tend, it will, probably, on all hands be allowed that it has not been unproductive in bringing more prominently forward many of the most interesting facts and conclusions respecting the structure and conditions of the heavenly bodies, and at the very least putting the public mind more fully in possession of those data which are necessary for carrying out any more imaginative speculations on reasonable grounds.

The question a point for philosophical conjecture.

In the former Essay* reference has been made to the nature and grounds of *Philosophical Conjecture*, and the place which it may legitimately and usefully

---

* See Essay I. § iii.

hold among the speculations of science. Those remarks may perhaps find an application in reference to such questions as that now before us.

Viewed simply as a question of philosophical conjecture, or rational probability, without reference to any ulterior consideration, the argument must be based on an extension of *inductive analogies*, a generalisation (so far as we can legitimately pursue it) upon the acknowledged relations of animated existence with physical conditions and cosmical arrangements adapted to it. And it is in this point of view that we must, in the first instance, proceed to consider it. At the same time, so numerous are the points of relation between the simple question of probability as to the fact of inhabitants in the heavenly bodies, and various collateral topics of higher interest, that the larger portion of the discussion, as taken up by the disputants already referred to, is in fact chiefly occupied by these collateral topics, to which the more simple question is manifestly regarded as subordinate : and it is probable that the public has been induced to feel an interest in the subject more from a reference to such ulterior considerations than

from the intrinsic attractions of the primary question
itself.

Review of
the argu-
ment ne-
cessary. But notwithstanding the ample discussion which
this subject has received from two such eminent disputants, it is still, in my opinion, left by them in an
unsatisfactory state.   Not so much in regard to the
mere question itself, which ever must remain in uncertainty, as with respect to a just appreciation of
the true grounds on which the discussion of it should
be taken up, as well as of the bearing and influence
which it may have upon those higher contemplations
with which both these writers (though in opposite
ways) have combined it.   It is, then, in this general
point of view, and in its connexion with other topics
of philosophical inquiry, rather than as to the mere
details of astronomy, that it is here proposed to treat
the subject.   Yet to a few of those details some
attention must be paid in the first instance.

Connexion
with past
history of
the world. The question as to the probable *present* habitable
condition of the planets or the existence of intelligent beings upon them, is closely connected with
the *past* history of the system.   And the discussion
of inhabited worlds has been justly much mixed up
with that of the process of transitions through which

they may have passed from an original nebulous or vaporous state to their existing condition.

The "nebular theory," as it is termed, of the origin of *our planetary system* is totally distinct from the phenomena of the *sidereal nebulæ* with which, nevertheless, it is often confounded. Though when the former theory was broached by Laplace it indeed received some confirmation by analogy from the discoveries made as to the varied forms of the sidereal nebulæ by the elder Herschel.

That highly distinguished astronomer had observed with his powerful reflectors, and minutely described, the forms and characters of a great number of those nebulæ, and thus there were supposed to be furnished so many actually existing instances of what had as yet been only a hypothetical speculation. Nebulous matter had been assumed to have existed in the solar system ; here there seemed cases of such matter really existing in the sidereal heavens. Moreover, there appeared to be great diversities of form and species of such matter. Some nebulæ presented the appearance of mere faint patches of dull light : others exhibited something like a nucleus, or brighter centre : others a distinct star surrounded with

The nebular theory.

Sidereal nebulæ.

nebulous matter : others various combinations of
stellar and nebulous appearances, often presenting
irregular portions, and many of variously formed
globular or elliptical structure.   Hence he was
naturally led to the conjecture that these might be
only gradual and *progressive stages of formation* from
mere diffused nebulous matter up to condensed stars,
or solid masses.

Sidereal
clusters.

And the analogy was further carried out when,
besides the nebulæ properly so called, there were
found a number of other bodies which to lower
powers appeared like nebulæ, yet with telescopes of
higher capacity *were resolved into clusters of separate
stars.*   This, then, seemed to be the ultimate stage of
their progressive evolution, originating out of mere
diffuse cosmical nebulosity, by degrees condensed
towards various centres, and at length absorbing all
the matter into distinct bright stellar bodies clustered
together—that is, clustered *to our eyes,* but really
at amazing distances from each other—forming vast
but separate groups in the heavens : of which groups
the whole of the fixed stars (ordinarily so termed as
distinct from the clusters and nebulæ), extending
laterally, and seen in thick perspective to form the

Milky Way, are believed to constitute only one, perhaps very subordinate, component part or cluster;
all the members of which are self-luminous, or
suns; one among which is our own sun, with his
little attendant planetary system, as invisible to
them as any systems they may possibly possess must
be to us.

We might well stop to expatiate on the magni-
ficent scene thus presented to our contemplation, as
both of the writers referred to have done with so
much eloquent effect.   But to return to the immediate argument; it becomes important (with reference
to a common confusion of ideas) to dwell upon the
consideration that the analogy of the sun and his
system is *not* with *clusters* or nebulæ, but with *their*
integrant stars.   And those who expect either to
confirm or to refute any supposed relations or conditions of our planetary system with what may be
observed or imagined in the sidereal nebulæ are
altogether on a wrong course.

Now the nebulæ, as originally observed in the
northern hemisphere by the elder, and in the southern by the younger Herschel, have been since, to a
considerable extent re-examined in the northern

Relation of
nebulæ to
the solar
system.

Resolution
of nebulæ
l y powerful
telescopes.

with the vastly more powerful telescope of Lord
Rosse; and, as the former observations had greatly
altered our conceptions of such nebulæ as had been
previously examined by Messier and others with
very inferior powers, and had resolved into separate
stars some which had previously been regarded as
purely nebulous, so the more gigantic powers now
applied not only presented many of the familiar
nebulæ under aspects in which their old features
could hardly be recognised, but succeeded in *resolving* into *clusters* of *stars* many objects before deemed
incapable of such resolution.   And it may be well
here to observe, that when the author of " The Plurality " describes these resolutions as into *patches* and
dots of light, seeming (if I understand him rightly)
rather to discountenance the notion of their strictly
stellar nature, in point of fact, I am able to state, on
the authority of those who have actually seen them in
Lord Rosse's instrument, that the appearance is perfectly and brilliantly that of *stars;* distinct effulgent
points of no sensible magnitude, and of whose stellar
nature no doubt could remain on the mind of the
observer.

Here, then, an argument has been raised : — As successive portions of nebulæ become resolvable when we apply successively higher powers, so it is inferred we may reasonably extend this argument, and fairly expect that this would always continue, as still higher powers might come to be employed ; and thus there would be no limit, and, in fact, all sidereal nebulous appearances would thus be shown to be most probably nothing but starry clusters, appearing nebulous simply in consequence of their enormous distance.   Hence, it is said all theories founded on such conceptions of nebulous matter must be given up ; and hence the opponents of such theories have enjoyed a triumph in which they have been often prone to indulge feelings not apparently much in character with so purely abstract and hypo-thetical a speculation, on which, nevertheless, opi-nions on either side have been maintained with a de-gree of vehemence little to have been expected from the nature of the subject.   To this, however, there is opposed one remarkable fact — viz., that in all the instances examined by these extremely high powers, wherever parts of nebulæ not before resolved have

Whether any limit to such re-solution?

been so resolved, there have been also brought to
light numberless new portions of the same mass of a
character apparently quite nebulous, of extreme
tenuity, bending about in delicate films in the most
capricious shapes.   It has hence been an inference
on the other side, that with widely ramified stellar
clusters there are always associated large masses of
yet unformed cosmical matter.

Some nebu-
læ real: not
merely op-
tical.
This perpetual disclosure of new unformed nebulous filaments and appendages to central stellar
clusters has been associated by the author of " The
Plurality " with some other considerations arising
out of the phenomena of certain other nebulous appearances in various parts of the heavens, especially
those singular and well-known bodies the " Magel-
lanic Clouds," so minutely and graphically described by Sir J. Herschel in his " Observations
in the Southern Hemisphere."   These present, in
remarkable juxtaposition, examples of almost every
form of stellar and nebular phenomena.   All which
the author considers as supporting the conclusion,
that these nebulous appearances are actually and
properly such, and differ from clusters of stars, not
merely in semblance from the optical effect of dis-

tance, but in their own nature as real aggregations of diffuse cosmical matter of some kind.*

As to the probable distances of this class of nebulæ, it has been observed, that the truly nebulous portions are evidently *physically* connected with the bright isolated stars scattered among them ; and these we have no reason to suppose more distant, on the average, than other stars. Lord Rosse, in his address to the Royal Society, 1853, observes, "In certain nebulæ, stars are so peculiarly situated that we can scarcely doubt their connexion with the nebular system in which we see them, and some of these stars are as bright as some of the stars known to be physically double ; as bright even as some of the stars which the latest Pulkowa observations have shown to have sensible parallax, and whose distance is therefore approximately known." This agrees with the argument of Sir W. Herschel †, on the principle of the visibility of stars in more powerful telescopes, being a measure of their distance, he calculates that, as stars are visible to the naked eye up to a certain order of magni-

Probable distances of some nebulæ.

* Essay on the Plurality, &c., p. 118.
† Phil. Trans. 1818.

tudes, that is, distances, so telescopes bringing
into view more orders, have their space-penetrating
powers numerically assigned on that scale. He observed, with the highest powers, a star which was
itself of the twelfth magnitude, still surrounded
by a nebulous haze. If the nebular part really consists of stars, only not separable by the telescope
from the effect of distance, then the central star
must be of such enormous magnitude, in comparison,
as would be at variance with all analogy ; and hence
he infers that the appendage is truly nebulous.

The great nebula in Orion remained unresolved
by all telescopes, even Lord Rosse's smaller one ;
and when, in 1846, the large reflector was applied,
though much of it was resolved, or gave appearance
of " resolvability," yet it would seem that some
portions still retained a nebulous appearance.*

Such considerations seem to indicate a probability
that some of these really nebulous masses may be
not more remote than some of the single stars which
compose our own cluster.

Some real test would probably be supplied, if

---

* See a Short Letter addressed by Lord Rosse to Prof. Nichol, published in his " System of the World," p. 55.

any of these nebular bodies should be found to possess *proper motion,* such as so many of the fixed stars have been ascertained to possess, and which stand out as residual phenomena, after all deduction for the ordinary corrections, and have been shown principally to arise from the real motion of the solar system in space. Hence the existence or amount of *proper motion* is, generally speaking, something like a measure of *distance.* If then, other clusters and nebulæ, not connected with single stars, should show considerable proper motion of such a kind as accords with the motion of the solar system, it would be a proof of their proximity. But until such proof has been given, it is obviously a premature generalisation to assert of all the nebulæ and clusters *universally,* that they are not more remote than the stars, because *some* of them may be so. If the theory, that *all* nebulæ are only remote clusters is to be abandoned, it can only legitimately be exchanged for the assertion, that we must distinguish nebulæ into two classes, those which are comparatively *near,* and are *truly* nebulous, and those which are *remote* and only *apparently* so.

Whether the parts of clusters and nebulæ may be

Question as
to revolu-
tions of ne-
bulæ about
their
centres. in motion round any point, is a question which has
not hitherto been much inquired into.   Some theoretical remarks have been thrown out by Sir J.
Herschel*, and some doubts expressed by others
as to the possibility of their preserving equilibrium
without a rotatory motion, or as to the possibility
of the law of the inverse square of the distance
being sufficient to account for their phenomena.
It is, however, obvious, that if there were the '
most rapid revolution in any portions of these
systems, they may be at too immense a distance to
enable us to detect it till after centuries of observation.   The sail of a windmill or a railway
carriage sweeps past the spectator close to it with
lightning speed ; seen at the distance of some miles,
it seems to revolve with extreme slowness, or to
creep along, in proportion to the diminished *angular*
visual space passed through in the same time.   A
distant planet moving with the velocity of several
thousand miles in a minute, presents no sensible
motion even to the astronomer, with the nicest
instruments, except by comparison of observations

---

* See Outlines, p. 866.

after considerable intervals. Double stars describe enormous orbits with velocities of proportionate amount, whose motions are not to be detected but by the most delicate measures at distant epochs. How much more, then, may the infinitely more remote components of a starry cluster seem to be at rest even though they may really be whirled in a second through inconceivably vast regions of space about their centre of gravity?

But there is a still more striking point which has been duly commented upon by both the writers before us — that singular feature pervading so many of Lord Rosse's nebulæ, that they appear in the form *of spiral convolutions of filmy nebulous matter, tending to a central nucleus or star.* And an analogy is immediately suggested with a revolving *system,* — with bodies urged towards a centre in contracting spirals, because the orbits they would have described round that centre are continually compressed by the action of a *dense retarding medium* through which they move. *Spiral forms of nebulæ.*

Lord Rosse* distinctly states his belief in the probability of *motion* in the spiral nebulæ : — " If *Probable revolutions in spirals.*

---

* Address to the Royal Society, 1853, p. 7.

we see a system with a distinct spiral arrangement, all analogy leads us to conclude that there has been motion; and that, if there has been motion, it still continues. The apparent motion is probably very slow, owing to the immense distance of the nebulæ."*

The filmy spirals of the nebulæ *may* doubtless be long strings and series of separate stellar bodies, all performing revolutions, whether in a retarding medium, or by virtue of the action of central forces acting by other laws than the inverse square; such, for instance, as the inverse *cube*, which, as is well known, Newton proved would cause a revolution in spirals. Our sun, with his attendant system, has been shown to be in motion in space; and probably revolving round a point assignable among the stars, and which some have fixed in or near the Pleiades. But if in *our system* there be supposed any analogy with the spiral nebulæ, it

---

* The spiral forms of Lord Rosse's nebulæ have suggested some theoretical considerations on their probable cause, as originating out of the rush of nebulous matter to a centre, which, unless (most improbably) in exactly opposite directions, would produce rotation : in fact, the same general idea as in Laplace's nebular theory; and just as water in a basin, if allowed to escape through a hole in the bottom, acquires a rotatory motion. These ideas have been acutely advocated by Mr. Nasmyth, in a communication to the Astronomical Society. (Notices, Vol. xv. No. 8. p. 220.)

should be remembered it is *not* with the motions of our planets or of Encke's comet and the retarding medium which it discloses, but with the cosmical revolution of our *whole system,* as part of *a cluster :* and to make out the analogy, it would be necessary to inquire whether it may possibly be the case, that other members of the solar cluster besides our planetary system, partake in a *sidereal* motion, and whether our milky way and little stellar group may be seen by the astronomers in *other clusters,* assuming the form of spiral wreaths wheeling round the Pleiades, and destined ultimately there to condense, and console the survivors for their lost sister.

The language employed in some parts of the speculations now adverted to, seems rather of a nature calculated to impose limits on our ideas of the immensity of creation, and thus to check one of the most sublime contemplations in which science has hitherto given us the privilege of indulging. The nebulæ can no more properly be called the " outskirts of creation," than the nearest planets. " What is there on the other side of the stars ? " is a question which a child may ask, but which the philosopher can only answer by alleging the fact of increasing

Too limited views of the extent of the sidereal heavens.

P

numbers of sidereal masses becoming successively
visible as our telescopes penetrate further into
space. To take up a theory which tempts us to
set bounds on the extent to which new worlds may
become visible to us if our telescopic powers were
conformably increased, would be to go in direct opposition to those suggestions to which all analogy
points ; to limit the immensity of creation, and to
attempt to model the plan of the universe to our
own narrow dogmas. A microscopic animalcule on
the sea shore might as well conclude that the
whole universe consisted of nothing but those
enormous masses, the grains of sand, which are all
that he could see around him, and as well deny the
existence of the land and the ocean, with their varied
inhabitants, because beyond his limited vision.

The speculations in which the true philosopher
may indulge on such a subject, guarded as they
ought to be by a becoming caution, will yet tend
continually to the increasing conviction at once of
the boundless extent of the universe, and of the
order and harmony which he is assured must pervade
every part of it.

But in the present instance the question of

nebulæ and the nebular hypothesis has been taken up not simply on its own merits, but as bearing, or supposed to bear, on the question of the *habitable condition* of these cosmical bodies or regions.

The resolvability of all nebulæ into separate stars and worlds is upheld as favourable to the notion of abodes for rational beings : and the existence of nebulous matter is discountenanced by those who are anxious to people every point of the resolved masses with animated existence. While, on the other hand, those who would pronounce the whole material world to be a vast and eternal " waste," excepting one small and favoured speck, find the hypothesis of nebular expansion through many enormous tracts of space more congenial to their ideas.

But it is difficult to perceive any very close or necessary connexion between the two questions. If, indeed, true nebulous matter be shown to exist in some instances, the real " star-dust," or " world-mist " of Humboldt, it certainly would not seem a very suitable dwelling-place for any kind of organised inhabitants : yet this might be merely because such portions of the system would not be as yet in

a stage of development in which we could expect
or imagine life to have begun.   But if this were true
in *some* parts of the universe, it would not
impugn the existence of inhabitants in *other* portions
of the system: it would merely show that different
parts were in different stages of progress.

In more advanced stages of consolidation, the
progress of life or the preparation for it *might*
be going on; and in regularly formed worlds
long ago completed, it *might* exist in full development notwithstanding that in other parts of the
system as yet in a state of nebulosity, we might
doubt or deny its existence.   If we choose to argue
the question whether a consolidated globe be a
necessary condition for the origination of life in any
form, but more especially of its higher forms, and
whether intellectual and moral faculties are to be
restricted to higher forms of organisation, or to any,
these are distinct inquiries.

The fixed
stars.

But the fixed stars properly so called — the members of our own cluster—some of which are revolving systems, binary, ternary or even multiple; —
the distances of some of which are actually within
the reach of our measures, and in comparison with

whose distances the diameter of the earth's orbit is
not an *absolutely* insensible point — are they inhabited? They consist of matter kindred with that
of our world; they are subject to the marvellous
affinity of the force of gravitation; they emit light,
that same agent which effects our eyes, and which
some terrestrial bodies emit. Whether we admit the
inconceivable idea of a universal ether in which these
remote bodies excite vibrations capable of ultimately
reaching us, or the equally or more inconceivable
idea of material molecules darted forth from these
same bodies and actually projected into our eyes —
and *one* of the two we *must* admit — these and many
more considerations show their kindred nature. But
being self-luminous, are they not suns? Must they
not be, like our sun, the sources of heat also, and
therefore probably intensely hot? If so, can they
be inhabited?

That the stars are self-luminous, and therefore (it Self-lumi-
might be argued) probably consist of matter in a state nous, or
suns.
of incandescence or combustion of some kind, is the
strongest fact against their being inhabited. This
however is no more than the same argument relative
to the sun, and, like it, susceptible, of the same

answer from conjecture as to the existence of an exterior photosphere. But this will be considered in the sequel.

It must also be recollected that the revolutions of binary or multiple stars are not analogous to those of our planetary system, but are the revolutions of two or more suns with their attendant systems (if any) about their centre of gravity.

Revolving systems of double stars.
The orbits of some of the double stars are indeed not greater than those of the exterior planets of our system, though their periods are rather greater (evincing consequently a less attraction between the two stars, or, what is the same thing, a joint mass less than that of the sun); yet small as these orbits are compared with that which our sun is describing round the Pleiades (or whatever the central point may be), the analogy would seem rather to be with this last orbit than with our planetary orbits. Whether in these instances the two mutually revolving suns may be attended each with his system of planets is a point which, doubtless, no existing observations tend to clear up. But it is one which theory pronounces unlikely, on the ground of the too great proximity of the rival attractions of the two

suns, which would make the maintenance of such planetary systems impossible, unless we suppose all the
members of each system to lie extremely close to its
sun.*   There is, however, no reason whatever why
they should not be thus constituted; nor why non-
luminous planets thus arranged should be without
inhabitants.

But even were the *double* stars suns, destitute of
planetary systems, there is no assignable reason why
the *single* stars (constituting so infinitely greater a
proportion) should not be thus attended: it being
obvious that no conceivable telescopic power could
ever discover such worlds to us: while some of the
nebulous stars might clearly present the rudiments of
such systems yet to be formed.

When from these remote systems we turn to that
with which we are more intimately connected, we
certainly recognise the existence of portions of *true
nebulous matter* in several parts of our planetary
world.   We find it in *comets* — in the *zodiacal luminous mass*, occupying nearly all the space within the
earth's orbit — in the *haze* surrounding more than

Nebulous portions of solar system.

---

* See Herschel's Outlines of Astronomy, p. 847.

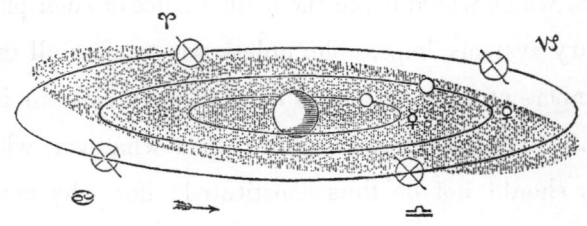

ZODIACAL LIGHT-MASS.

one of the small planets, and most probably in
those as yet little known forms of cosmical matter,
revolving under the influence of gravitation, which
give rise to meteors, and especially the periodical
star-showers.

To give some idea of the magnitude of such
masses it may suffice to mention Encke's comet,
which at a distance from the sun equal to that of the

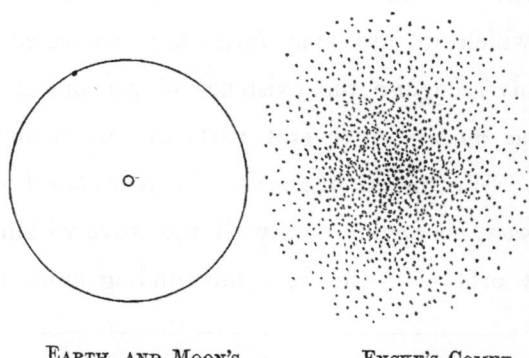

EARTH, AND MOON'S   ENCKE'S COMET.
     ORBIT.   (Relative Size.)

earth, occupied with its nebulosity a sphere greater than that of the *whole* moon's orbit; and the comet of 1843, whose tail according to Boguslawski equalled in length the *entire distance from the sun to Jupiter,* or 494,000,000 miles, with an average breadth of perhaps 500,000 miles.   And when we further take into account the multitudes of comets, so great as to justify the assertion of Kepler, that the universe is full of them, we shall have little remaining doubt as to the plentiful existence of true nebulous matter. Indeed, the late Mr. Baily, than whom no man was less likely to be led astray by fanciful hypotheses, did not hesitate to speak of comets as being detached fragments and remains of the original nebulosity; which " want of mass has saved from the extreme condensation which the planets have experienced ; " as having been originally projected in parabolas and then, perhaps, from the resistance of the uncondensed remaining nebulous matter with which they were sur-rounded, compelled to revolve in limited and re-en-tering orbits, — and as forming " a link between the present and past states of our system, showing in their obedience to the law of gravitation a pre-sumption that that law has been unaltered since the system was nebulous, and now holds good in the nebulæ which remain unchanged." *

Nebular
origin of
our system.

The existence of such unformed matter adhering to our system has been appealed to as at least affording presumptive evidence bearing on the probable origin of that system ; and while one party has been led to support the nebular theory of Laplace, perhaps to an undue extent, and to find in it a solution of the entire creation of our system, in some instances carried out by attempts at even *numerical* verification, which have been shown to be erroneous,—another party has with equally unphilosophical vehemence and warmth denounced that theory as replete with every form of mischievous error, and has indulged in misplaced feelings of triumph when they fancied they saw its downfall in the detected errors of M. Comte's computations, or the resolution of the sidereal nebulæ in Lord Rosse's telescope.

It is no part of the present object to assert or to defend the nebular theory, except on the general ground that it is a perfectly *legitimate* kind of *con-*

---

* Address to the Royal Astron. Society, 1837.   Notices, vol. v. p. 51.

jecture.   One objection, however, may be noticed in An objection answered. passing, because it has been sometimes referred to as destructive of that theory.   It arises, however, from a mere oversight.   A body *rotating on its axis* in a dense medium will have its motion simply *retarded;* a body *revolving* freely in an *orbit* in the same medium will have its *orbit contracted* and its *motion* consequently *accelerated.*   This distinction has apparently been lost sight of by some who raise objections to the nebular theory, on the ground that the times of the rotation of the sun and of the revolutions of the planets are so different.   They must necessarily become so under the conditions supposed, though originally one rotating mass.   As soon as a planet was separated, it was transferred from the dominion of the one law to that of the other.

The nebular theory of the solar system, soberly understood, is a philosophical conception worthy of the subject which it illustrates : starting from the fact of central heat in the earth, and the indications of it in the spherical forms of the other planets, we are unavoidably carried back to a period when all was in fusion ; and thence to a period when all was vapour or nebulosity, out of which by successive

Rational claims of the nebular theory.

cooling and condensation, and the rotatory motions ensuing upon the rush of matter towards the centre, the existing system *may* have been developed according to regular and uniform laws; and which is so far a rational and consistent *conjecture* (for it can be no more), eminently conformable to the grand principle of cosmical unity and order. The truly philosophical advocate of such a theory, following the track of inductive analogy, might not be disposed to assign organised inhabitants to any of the bodies so formed, till after immense periods of cooling and consolidation. But he would be led into no dogmatising on the subject, and would simply call on us to be guided by the analogies suggested by what we know, subject to the condition that in the infinity of what we do not know, equally grand principles of order and unity must prevail : principles and laws not necessarily the same as those with which we are acquainted, yet equally invariable under the conditions to which they are adjusted.

Common origin of all planetary matter.

But if the common origin of the planets and the sun, from one primary nebulous mass, be admitted, this further consideration is forced upon us; viz., that as they were thus all parts of the same material mass, that mass must have contained, mixed up in it, all the elements of every possible product of nature, organic or inorganic, and the germs of all vitality, even to its *highest* forms, *in so far* as they partake of an animal nature; and we may therefore suppose in all the planets the same inherent capacity for having life evolved in them from its lowest up to its highest forms.

If from the conjectural past history of our system we proceed to consider its *actual* condition in reference to the question of inhabitants in the various bodies composing it, we must look carefully to facts; and by the writers who have treated the subject on either side, we are sent back to our works on astronomy, to the ascertained data of observation, and the inferences from them, to consider the magnitudes and distances of the planets, the proportions of light and heat they receive, the variations of their seasons and lengths of their years; the satellites furnished to so many of them, and the rings to one; we are shown well-known calculations of the force of gravity at their surfaces, the weights with which beings placed there would be pressed, the known density of the materials of which they are composed, the

Present condition of the planetary system.

presence or absence of atmospheres; the former
being admitted, on all hands, in every planet except
our moon, and even there still a question with
some.   From all these and the like data, fair conclusions may be drawn as to their *capability* of
sustaining organised inhabitants, and thus we may
be legitimately led to *conjectures* as to the probability of rational or moral beings tenanting their
surfaces.

Diversity in inhabit-ants, if any.

Whatever may be the opinion entertained on these
points, it will be on all hands admitted that we must
suppose, at the least, great diversity in the nature
of the possible inhabitants of the different planets,
corresponding to the known diversities of conditions
subsisting in them.   But the inquiry mainly refers
to the question, whether these differences must be so
great as to preclude all idea of a nature *analogous* to
our own, in a physical point of view; it would be of
course easy to grant metaphysical entities or spiritual
existences of an unknown kind, wholly different;
but this is not the question before us.

We are acquainted with moral and intellectual
life in finite beings only as connected with a material
organisation; we see our own world suited to be the

dwelling-place of such organisation; we see other
worlds around us presenting many external points of
analogy — are they suited to be the dwelling-places
of beings in any way *analogous*? The question rather
is, whether there are any positive grounds for supposing the diversities so great as to destroy all idea
of any common or kindred nature in such inhabitants, and thus practically to put them so entirely
out of the category of beings within any range of
sympathy or connexion with ourselves, as would be
virtually to deny their existence.

The object of this essay is not controversial; it is   Difficulties
to be fairly
not, therefore, intended to go into the questions of   considered.
detail raised by either of the disputants on the subject of the habitable nature of the planets. It is
my object to look at this question rather in its
more general aspect, than as referring to particular
planets, and in relation to the broader argument
applying to all. There are, however (as we have
already noticed), a variety of considerations to be
taken into account; with all analogy in favour of
the *possibility* of inhabitants in the planets, it is still
little more than a possibility. If there were no
arguments of another kind to oppose, it might amount

to *probability*.   But there are some such opposing arguments, which must be carefully noticed, such as those derived from the state of our moon, the geological history of our earth, and the condition of the sun, the asteroids and comets, to all which we shall refer in due order.

The sun.

As to the sun himself: the ingenious speculations of the elder Herschel and Arago have assigned him a race of inhabitants very like ourselves, living upon the solid globe or nucleus, over which is spread the resplendent atmosphere or photosphere, the source of those rays which convey both light and heat, and which, it is conceived, may emanate from the envelope without affecting the central body.

Nature of the solar heat.

Indeed, on this point there is one consideration often not sufficiently attended to.   The solar heat is entirely of a peculiar nature, unlike that which emanates from a terrestrial hot body simply cooling or radiating its heat.   The solar heat is not derived from the mere cooling of the sun, but is conveyed, as it were, *in* the rays of light, as a *vehicle*, and *never* becomes *sensible as heat* till the *light* is absorbed.   It is, therefore, probable that these rays may owe their extrication from the sun to *some other cause than*

elevation of temperature. It is an effect elicited or
produced by the action of certain rays, which are no
more properly rays of heat than a galvanic current
can be called a current of heat, because, when
stopped, it excites heat. The solar rays pass freely
not only through empty space, but even through
air and all *transparent* media, *without heating* them;
they never excite heat till they are impeded by a
solid, or at least an opaque body.

" The temperature of space " is a term which some
philosophers use, probably meaning the proper temperature of *some medium* diffused through the celestial
spaces, and which is independent of the sun's radiation.
This proper temperature is supposed by some to be
extremely low; at any rate, it manifestly depends on
the degree of this temperature to what extent any
planet shall *retain* the heat imparted to it by the
sun, and the loss of heat will be greatly modified
in proportion as their surfaces possess high or low
radiating powers.

The time of rotation of the planets, again, is a ma- Temperature of the
terial element in modifying the degree of heat they planets.
receive from the sun, from the comparative rapidity
with which points on their surfaces pass from under

Q

the heating rays of the sun. But this, again, must
be greatly modified by the individual temperatures
of the planets arising from internal heat, the remains of their primeval high temperature, and dependent on their rate of cooling, which may be
different in the different planets, from a state of
primeval fusion. Some rough calculation, perhaps,
might be instituted from their known densities and
magnitudes which might give an idea as to the
possible relative temperatures, or states of consolidation, which they may at the present time have
simultaneously reached, supposing them to have
commenced from a common period of fusion.

On such points, indeed, we have few data; but
they are clearly essential to be taken into account
before we can pronounce on the actual temperature
of any planet; and even that, if known, can but
little affect any conclusion as to the existence
of inhabitants, since even with our own frames
we know what great differences of temperature can
be sustained without inconvenience ; even eno mous
differences have been endured without actual injury ;
and the conception of modified organisations to suit
any difference of temperature is by no means difficult.*

---

* On this subject the reader is referred to a valuable and elaborate

There is also this material consideration to be taken into account. Atmospheres of an aqueous nature have been supposed surrounding the planets.* The *solar rays* will penetrate through such an atmosphere in proportion to its *transparency ;* but when the solid body is heated, it radiates a new species of heat, which is *not* transmissible by radiation through the aqueous envelope. The planet might thus retain a large proportion of the heat acquired, or at any rate could only lose it by the heating of the supposed atmosphere, which may then externally radiate it away. Such an envelope might thus greatly modify the temperature of a planet.

Of the moon, from its proximity, we 'of course The moon. know more than of the rest of our system. The visible details of its physical conformation—or, as a recent writer somewhat strangely called it — the *geology* of the moon, are familiarly known, and lead

paper by W. Hopkins, Esq.) in the Cambridge Philos. Transactions, vol. ix. Part. IV., 1856. As to many details of the astronomical data respecting the planets, double stars, &c., numerous inaccuracies in the computations of the " Essay " are pointed out in an able and acute little tract, entitled " A few more Words on the Plurality of Worlds "; by W. S. Jacob, F. R. A. S., astronomer to the Hon. E. I. C. at Madras. London, 1855.

* See Brewster, p. 54.

to the general idea of a vast waste of extinct craters without traces of seas or rivers, without a perceptible atmosphere, unless it be one of the most contracted kind, not reaching so high as to the tops of the mountains.

The most delicate thermoscopes have failed to detect the slightest heat in the concentrated rays of the moon; that is, in the sun's light scattered and dispersed away as it is after reflection from her convex and rough surface.

The heat of temperature which the moon herself acquires by absorption of these rays, if it could be radiated in sensible intensity to so great a distance as to the earth, would probably be all absorbed by our atmosphere and clouds, if not by any other cosmical medium, before it reaches us.

Yet conjectures are not wanting on the other side: according to which the absence of water has been accounted for by its transfer from the surface, for a whole fortnight exposed to the sun, to the opposite side " by distillation *in vacuo*, after the manner of the cryophorus."*   These and other arguments must

* Herschel, Outlines of Astronomy, § 431.

at least leave the question fairly open to discussion. Again, with regard to the visible hemisphere of our satellite, it has been remarked, that if the lunarians are beings of at all similar social nature with ourselves, they would naturally inhabit cities; and large cities would be fully visible to powerful telescopes. Again, if they had extensively the use of fire, and manufactures dependent on its use, this would hardly fail to reveal itself by smoke. From the absence of such indications, then, it is affirmed that they have not anything resembling our manufacturing towns.

But against this it has been very recently contended *, on the strength of analogous actual observations, that at such a distance cities and smoke would not reflect light so as to be distinguishable.

At any rate we see enough to assure us that the inhabitants, if any, must be very differently constituted from ourselves, to live in enjoyment under the manifest singularities of climate and physical condition and the extreme alternations of temperature to which they must be subject. But geology is not without its use in hinting at some kind of

Lunarians necessarily very different from terrestrials.

---

* See a Paper by Prof. C. P. Smyth in the Edinb. Phil. Journal, 1855.

analogy in the moon with stages of formation which our earth has passed through. When the lunar volcanoes were active, the elements of water must have been present; and the terrestrial water, and even the atmosphere, must have been in a very invisible state when the earth was at an extremely high temperature. It may be added, that Mädler and Beer, after their accurate and elaborate examination of the physical structure of the moon, conclude with a distinct speculation on the probability of its containing inhabitants, however differently constituted from ourselves.

But all the foregoing speculations assume that the hemisphere of the moon, which is towards us, is similarly conditioned to the rest of its surface. Now, even at a very recent date, some results have been obtained of a highly curious nature with respect to the moon, which materially affect the question, inasmuch as they show a peculiarity in the *visible* hemisphere.

Professor Hansen * has lately pursued some elaborate computations of certain corrections necessary

---

* Astronomical Society's Notices, vol. xv. p. 14.

to be applied to the inequalities of the moon's motion, deduced on the theory of gravitation, and thus involving as an element the attractive force of the moon, dependent not only on its absolute mass, but also on its figure. Hence he arrives at the theoretical conclusion that it is necessary to suppose *the centre of gravity of the moon different from the centre of its figure, and farther from the earth than the latter;* in other words, the hemisphere of the moon turned towards us is *more* raised above the *mean level,* as referred to the centre of gravity, than that away from us; a condition to which its ocean and atmosphere, if it had any, would conform: this relative elevation at the middle point of the hemisphere amounts to nearly twenty-nine miles. And the summit of a mountain or table land of that height would be necessarily destitute of water or atmosphere; though both may exist with all the attendant phenomena of life in the hemisphere away from us, which is not thus raised; and even at the boundary some traces of an atmosphere might be perceptible.

The intensity of the solar rays, which would occasion so hot a climate in Mercury, may be greatly mitigated by the density of the medium by which he

Mercury, Venus, Mars.

is surrounded; and the temperature to which the inhabitants of any part of Venus may be exposed will necessarily be very variable, from the extreme inclination of her axis, which gives rise to two winters and two summers in the course of her short year to the greater part of her surface, as is the case on a less notable scale within our own tropics, her tropics being near her poles.*

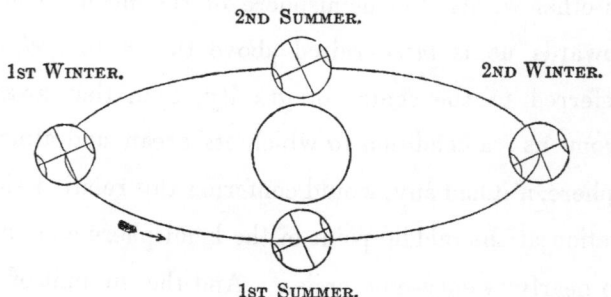

2ND SUMMER.

1ST WINTER.   2ND WINTER.

1ST SUMMER.

SEASONS IN VENUS.   Inclin. 75°.

The planet Mars is on all hands admitted to be circumstanced so similarly to our earth, that little discussion can be needed.

Planetoids.   The kind of creatures capable of inhabiting the

* On this point (as, indeed, on most others of interest connected with the physical facts of the planetary worlds), the reader will perhaps hardly need to be formally referred to Admiral Smyth's "Cycle of Celestial Objects," vol. i. p. 107., &c.

asteroids, or planetoids as they are now more properly called, is a question on which Sir J. Herschel *
has not thought it unfit to offer some conjectures.
But if they are uninhabitable — if they are exceptional in this respect — so they are in all others.
Their number, so far from being exhausted by
modern discoveries, will very probably be found to
be infinite, and of all sizes, from the first known,
downwards; not fragments of an explosion (which
savours of the convulsionary hypothesis), but globules
or molecules, gradually condensed *from a remaining
ring of cosmical matter*, other parts of which *may*
even yet be destined to undergo further condensa-
tions and combinations, and to form larger planets,
which may not for myriads of ages be approaching
the stage of bearing inhabitants.

The materials of which Jupiter is composed are of Jupiter and
the superior
a specific gravity about equal to that of water, which planets.
is the same nearly as that of the sun.   The essayist,
in his assumed magisterial vein, lays it down as by
no means an arbitrary hypothesis, that Jupiter is a
globe of water; and argues accordingly that his in-

---

* Outlines, § 525.

habitants, if any, must be aquarian creatures of a soft, pulpy, boneless, watery character, to which, he thinks, we should naturally feel it very difficult to ascribe intelligence or moral attributes; that is, without violating those analogies which we are so prone to form (perhaps groundlessly) from contemplating our own species.

But, as Sir D. Brewster, on the other hand, very justly observes, there are many solid substances, and even some minerals, as pumice, pitchstone, &c., and the metals of the alkalis, of less specific gravity than water. Jupiter, therefore, may just as well be composed of solid materials, and be tenanted by animals capable of living on land, as by aquatics. Similar calculations have shown, that in all the outer planets the conditions of gravitation are nearly the same; nor need the small specific gravity requisite for such animated beings occasion any difficulty. On our own planet, animals differ widely in this respect. It is hardly necessary to remark that birds, *e. g.*, have their bones, coverings, &c., of much greater specific lightness than the corresponding parts of terrestrial animals.

Nay, Sir D. Brewster has shown, by direct calcu-

lation, that even human beings, constituted as we
are, would not really be much inconvenienced if
transported to the surface of Jupiter; that buildings
and trees, such as we have on our earth, might stand
and grow securely in so far as the force of gravity is
concerned : and the same would be true for the
planets exterior to him.*   At any rate, when we
reflect on the extremely varied forms of animated
life on our own globe, on the diversified structures
of different classes of animals, and the marvellous
adaptations of their respiratory and circulatory
functions to the conditions of their existence under
the most varied circumstances, yet all preserving the
most recondite relations to analogy and unity of composition, we conceive there can exist no difficulty
in *imagining the possibility* of living beings constructed
with bodies of greater or less specific gravity, suited
to the most widely different conditions of gravitation
or atmospheric pressure in which they might be
destined to live, and with respiratory, muscular, di-
gestive, or locomotive powers and capacities developed
in infinitely varied degrees, according to the different

* Brewster, p. 64.

conditions under which they might subsist, and the media in which they might have to move — yet always preserving an unbroken analogy with *some* grand and universal scheme of uniformity, of which we enjoy only partial glimpses; while under any such variety of external form or condition, they *may* be equally capable with ourselves of being the recipients of higher principles of intellectual, moral, or spiritual life.

Argument from the reflected light of the planets.

One of the arguments respecting the physical condition of the heavenly bodies, has been ingeniously derived from the polarised condition of the light reflected from their surfaces. In our own globe, this modification exists in the blue light of the sky. Sir D. Brewster, in referring to this fact, alludes to conjectures which assign, on this ground, an *aqueous* nature to the medium reflecting that blue light; and assuming that the fact is so, he goes on to apply the remark by observing that, to a distant spectator, the polarised condition of the light reflected from the earth would be a proof of the *aqueous* nature of its envelope, and hence such polarisation, observed as it has been in the light of the planets, is a like

proof of the existence of *water* in them.*   This inference, however, appears to me unfounded : all transparent bodies, not only water, polarise light by reflection : and some opaque and rough bodies, and even
metals (contrary to the assertion of many elementary
books) polarise some portion to a very sensible amount.
Hence the inference of an *aqueous* reflector is not
a necessary one.   In point of fact, the light of the
moon and of comets has been found to be polarised,
where the presence of water is more than doubtful.
If the planets presented plane surfaces, or if we
could otherwise determine the *angle* of polarisation,
then the inference as to water might be verified.

The condition of comets is perhaps in some re-   Comets.
spects as well understood as that of the solid planets,
notwithstanding their apparently more singular appearance, and, in some instances, enormously distant
wanderings.   They are certainly *transparent as
masses*, whether composed of gaseous matter, or
much more probably of minute solid molecules
loosely aggregated, and probably kept so by a high
state of electric tension, through whose interstices

---

* Brewster, p. 54.

light passes, but which may also be transparent
themselves. It may be too hasty an assumption to
assert that even in their present state they are not
likely to have organised inhabitants; yet they may
at some future period be destined to *condense*, as
Gambart's comet, separating into two, and each exhibiting a nucleus, shows a tendency to do; and may
in time become fixed and consolidated bodies of our
system, the future abodes of life.

Planetary
comets and
asteroids.

With respect to that singular system of comets
of short period, consisting of at least five or six
members, whose aphelia (or greatest distances)
lie all within a little beyond the orbit of Jupiter,
and which are all mere vast nebulous masses,
yet as much parts of our system as any of the
solid planets, it has been distinctly shown by
Leverrier that they have probably been once tra-
versing the depths of space in infinitely long orbits,
and have at length been fixed in our system by the
attraction of Jupiter. May they not, in the revolution
of ages, possibly be destined to become more solidified
members of our planetary world? Nor can an astronomical imagination divest itself of the idea of some

possible relation between this system and the ring of planetoids or scattered planetary globules, approaching in so many respects a cometary nature, which occupies the region at about the mean distance of the planeto-cometary bodies. In all these we may be at liberty to fancy stages of progress towards worlds, and that consequently in each there may exist at least the germs and seeds of organisation and life.

From what was before remarked as to the nature of the solar rays, arises this remarkable consideration with respect to comets, at variance with what is commonly supposed ; being, as we know they are, extremely *transparent,* however near they may approach the sun, his rays will *pass through them without heating them.* At least it can only be from *imperfect transparency,* which doubtless may be increased as they are more *condensed* at perihelion *, that any solar heat can affect them. But even then the effect may be much less than is commonly imagined. This consideration may not be without value in reference to the conceivable idea of minute organised beings,

---

* See Professor C. P. Smyth on Comets, Edin. Trans. 1850.

monads or animalculæ, peopling the fine molecules of which cometary matter may consist.

Conditions of planets differ only in degree from those of the earth.

In general, so far as anything is made to appear on certain and definite astronomical grounds, the whole question appears to be one of *degree*: the conditions of gravitation and atmospheric pressure, of heat and light, and the like, are only questions of *degree* in reference to all the planets of our system: they all resemble the earth *more or less*, they enjoy *degrees* of the *same* physical influences *more* or *less*. But their conditions in *no* instance differ *in kind*. From the brightest, hottest, or most intensely gravitating, up to the coldest, darkest, and most feebly attractive, there is but an enlarged or contracted *scale* of influences, and not a change in their kind or *nature*. So far as these considerations go, we can therefore see no proof or even presumption whatever against their possessing inhabitants; in structure and function, differing perhaps not more widely from those of the earth than the other conditions differ: — in the adaptation and development of their forms or faculties modified in *degree* to as great an amount as the conditions of their existence may be diversified; — yet without any such *essential* and *radical* diversity in *kind* or *nature* as would place them out of all analogy with what we know of at least one small portion or member of the *great whole, or kindred scheme of cosmos.*

The considerations furnished to us by geology from its disclosures of the past history of the formation of the earth's crust with its organised inhabitants are perhaps among the most material aids to any inferences as to the structure and destination of other planetary bodies. *Analogy from past state of the earth.*

And here one of the points most dwelt on is the assertion of the recent date of man on the globe, which has been commonly assumed to be settled on what is confessedly mere negative evidence. It is, indeed, an opinion at present current among geologists, that man cannot claim a higher date than a period *later* than the latest of the tertiary deposits; but how many millions of years ago was the latest of these deposits is not so easily settled. This opinion is grounded solely on the *mere absence* of any remains hitherto detected, and with no powerful analogies in support of the negation, but with every probability to the contrary afforded in the apparent fitness of the state of the earth for man being its tenant at a much earlier date than that commonly assigned to his *Earth long without man.*

origin. There seems, however, to be a peculiar fondness in some minds for triumphing in the assumed recent origin of the race, not very intelligible on *philosophic* grounds, and in which both writers in the present controversy seem more or less to partake.

It is, indeed, on all hands admitted that we neither have any evidence, nor would analogy allow us to expect any, of the existence of man (or, indeed, even of the higher mammalia throughout all the vast periods of the earlier formations.

But, after all, to what does the argument amount? Only to this : that the earth, during all these incalculably long past periods, was nevertheless in a state of *preparation* for man's habitation; and thus, although the planets may now be destitute of inhabitants, they may possibly be only undergoing similar changes anticipatory to a similar result; and thus, if the uninhabited state of the planets be admitted, we must still enlarge our ideas, we must embrace in our view not merely the *present* moment of time, but the whole expanse of the past and the future, and thus regard all worlds as equally *related*, actually or prospectively, to the destinies of moral and intelligent beings.

Thus such considerations as geology furnishes are not without their use in pointing to a very possible interpretation of many parts of the planetary economy. As our earth has undergone successive changes, and was probably for vastly long periods destitute of all organised life, or at least of all its higher forms ; so it may be fairly argued, if any of the planets present features incompatible with organised life at the *present time*, it by no means follows that they will not have life developed in them at *some* future time, even up to its highest grade.

Planets may be for a time devoid of inhabitants.

Yet afterwards become inhabited.

Or, again, others may possibly argue, they may have already attained that point, and have been since undergoing a retrograde influence of a destructive kind, reducing them to an uninhabited state. Yet surely, when all that we *know* of the past history of the universe, little as it is, is wholly stamped with the character of *advance*, we cannot easily reconcile ourselves to the idea of *retrogradation* or destructive agencies in any part or member of the system. But even were this so, it might be more philosophically regarded as merely one phase out of a vast series of changes in their recondite arrangement, to be succeeded again by renovated light and life.

In all that is urged relative to the physical conditions of the planets, there is nothing to show, that they may not experience changes in those conditions, or may not have already experienced them, in as great a degree as the earth, we know, has done, and beyond a doubt will continue to do : and this by the slow operation of immutably ordered and profoundly adjusted series of regular laws and physical causes. If the earth in the process of progressive evolution has at a certain stage become the scene and site of intellectual and moral existence, the other planets also may have been so, or may at a future time become so.

Probable general diffusion of life, present or future.

Looking at the subject solely as a question of plausible philosophic conjecture, and guided as we should be by the pure light of inductive analogy, all astronomical presumption, taking the truths of geology into account, seems to be in favour of progressive order, advancing from the inorganic to the organic, and from the insensible up to the intellectual and moral in all parts of the material world alike, though not necessarily in all at the same time or with the same rapidity; in some worlds one stage being reached, while in others only a comparatively small advance may have been made.

It is on an enlarged view of the uniformity pervading the entire system of the natural world, both in space and time, that the assertion of rational inhabitants in other worlds has been defended by Œrsted in that eloquent posthumous work already referred to.* And though he agrees so far with the author of the Essay as to admit the possible *existing* absence of inhabitants in many worlds, on the same principle as in the case of the earth for such innumerable ages prior to the introduction of man, yet it is only on the truly just and philosophic ground of supposing universal and perpetual change and progress according to some grand laws as yet unknown, of which the existing condition of things forms but one small subordinate stage and portion, and which everywhere tends to prepare the way for the higher development of rational and spiritual life.

Œrsted has in fact pursued this topic in more

Views of Œrsted.

---

* The Soul in Nature, &c.

R 3

than one part of his various essays, and on several
grounds.   In a portion of these reasonings he adopts
a tone of somewhat vague metaphysical speculation
in which I cannot altogether follow him.*   But so
far as the present discussion is concerned, and in a
more strictly physical point of view, the following
passage will present a fair idea of the nature of his
reasoning : —

" If we are now thoroughly convinced that everything in material existence is produced from similar
particles of matter, and by the same forces, and in
obedience to the same laws, we must also allow that
the planets have been formed according to the same
laws as our earth.   This we know, however, that
they have developed themselves during immeasurable
periods of time in a series of transformations, which
has also influenced the vegetable and animal creation
of those periods.   This development began with the
lower forms, and advanced by gradual steps to the
higher, till at length in the most recent periods a
creature was produced in which self-conscious knowledge was revealed.   We must, therefore, allow a

---

* See especially, pp. 53, 54.

similar mode of development in the other planets. There may be many which have not yet attained such a degree of development as our globe, or again other far higher beings may have been created: but everywhere the creatures endowed with reason are the productions of nature in the same sense as ourselves, that is, their understanding is bound up with the organs of their body; therefore the nature of their understanding cannot be fundamentally different from our own, but must obey the same laws." *

I will conclude with one more extract, which seems to complete Œrsted's view of the possible characters of the inhabitants of other worlds, and submit it without further comment to the judgment of the reader : —

" The variety in the nature of the planets of our own system is very great; but if we extend our thoughts over the whole universe, the differences are endless. On some planets the creatures may be possibly on a far larger scale, on others far smaller than on our own ; on some, perhaps, they are formed of less

---

* Ibid. p. 108.

solid matter, or may, indeed, approach the transparency of ether ; or, on others again, be formed of
much denser matter. The rational creatures on
some of the planets may be capable of receiving far
quicker, more acute, and more distinct impressions
than on the earth, and on others it may be quite the
contrary. If we now turn to the mental forces and
mental development, we cannot acknowledge less
variety. We may imagine that there are reasonable
beings with weaker faculties than our own; but if
we properly appreciate our present distance from
the aspirations of our reason, we feel compelled to
acknowledge that an endless nu...ber of degrees of
development may exist above the point we have
reached."*

---

* Ibid. p. 129.

§ II. — THE ARGUMENT CONSIDERED IN A THEOLOGICAL POINT OF VIEW.

Mutato nomine de te. — HOR.

THUS far I have endeavoured to restrict the view of the question to its *proper physical* and *philosophical ground — the extension of inductive analogies by reasonable conjecture.* But the writers on either side have not confined themselves to this view; they have in several respects availed themselves of *other grounds* of argument, and have had a special aim at ulterior and higher conclusions.

The discussion has extended beyond the physical question.

Radical de-
fect of both
views.

In the reference to such higher views, and the manner in which the question of planetary worlds is treated in its application to them, there seems much of a very questionable nature in the speculations on either side. But, in my opinion, the *radical fault* which pervades the whole discussion lies in the primary fact, that the question *is removed* from its proper *basis* of inductive conjecture and philosophical probability, and placed altogether on the new and unphilosophical ground of conformity to theological belief. The investigations throughout seem to be carried on, *not with a view to* PHILOSOPHICAL TRUTH, *but to serve an ulterior purpose;* a procedure which stamps the whole inquiry on either side with a character alien from scientific independence or freedom of inquiry.

Singular
representa-
tions
adopted in
the Essay.

In the first instance, we cannot but be struck by the somewhat singular tone in which the physical question of the structure and origin of the planetary system is spoken of by the author of the " Essay," under the license of his anonymous mask : — " The planets and stars are the lumps which have flown from the potter's wheel of the Great Maker, the shred coils of which in His working sprang from

His mighty lathe — the sparks which darted from
His awful anvil when the solar system lay incan-
descent thereon — the curls of vapour which rose
from the great cauldron of creation when its elements were separated." *   Again, the ideas broached
with respect to the arrangements of the planetary
system, and especially as affecting the rank and
position of the earth, are characterised by a similar
tone of paradox.   Thus the author observes — " The
earth's orbit is the temperate zone of the solar
system, and in that zone only is the play of hot and
cold, of moist and dry, possible."† . . . " The earth
is the largest planetary body in the solar system . .
the vast globes of Jupiter, Saturn, Uranus, and Neptune, which roll far above her, are still only huge
masses of cloud and vapour, water and air. . . . .
The earth is really the domestic hearth of this solar
system, adjusted between the hot and fiery haze on
one side, the cold and watery vapour on the other.
This region only is fit to be a domestic hearth, a seat
of habitation; and in this region is placed the largest
solid globe of our system; and on this globe, by

---

* Essay, p. 243.   † Ib. p. 196.

a series of creative operations entirely different from any of those which separate the solid from the vaporous, the cold from the hot, the moist from the dry, have been established in succession plants, animals, and man."*

Undue pre-eminence given to the earth.

Such representations seem somewhat to resemble the geography of the Chinese, where the Celestial Empire, stretched to gigantic proportions, occupies all the most central and fairest portion of the map, and the wretched inhabitants of Europe and America are condemned to its insignificant outskirts and remote corners.

But in the ideas thus broached it is not *mere* novelty and paradox on which I would animadvert. Such speculations appear like a *retrograde* movement. The author's masquerade assumes rather the fashionable *mediæval* costume, and affects a sort of pre-Raphaelite astronomy. His theories carry us back to the schemes in old books, with a gigantic earth, and a pygmy sun and planets performing their humble circuits round it. His speculations evince, if not a literal and physical, yet a moral *Ptolemaism :* they

---

* Essay, p. 203.

seem conceived in the spirit of the dark ages, which made man and his interests the sole centre and end of the universe, and would bend everything else in subserviency to them. The mediæval astronomy, as it made the earth the centre of the universe, so it helped to foster man's proud conceit, that he was himself the centre and sovereign of the moral creation. And this was intimately wound up with the superstitious belief of the age. The heavens revolved round the earth, to which everything was subservient; so man was the " microcosm," or emblem and type of the greater world without; and with him and his fortunes all nature was connected; to his benefit, or for his retribution, all things were made to be conducive. The planets shed their friendly or hostile influence at his birth. Their combinations and conjunctions were arranged to foretell his fate; and the return of comets and recurrence of eclipses were merely notices put forth to advertise nations and sovereigns of their approaching destiny.

So, again, the theology of the age took up the same ideas, and made them the strongholds of its dominion; and the supremacy of an exclusive creed and the paramount dignity of the church were deeply

The mediæval astronomy connected with theological views.

concerned in the maintenance of the central position
and immobility of the earth. As that position represented the metropolitan throne of the spiritual universe, which could in no way be consistent with a
location in a subordinate and revolving planet, an
insignificant globe among innumerable others — some
far superior; so the earth's immobility typified the
infallibility from which the authority of the church
emanated, and whose pretensions could not but be
placed in jeopardy when the former was assailed.

It should not be forgotten that the assertion of a
plurality of worlds was one of the heresies for which
Giordano Bruno was roasted alive; so that it be-
hoves Protestants to watch the inroads of any theological dictation of the opposite view, which the
tendency of the essayist's language seems to threaten.

Similar
spirit in
some of
these specu-
lations.

But though the same precise doctrines and pretensions of infallible authority may have now disappeared,
especially in Protestant countries, yet prepossessions
of a very kindred nature, and evincing exactly the
same *animus* and spirit, — " mutato nomine," — prevail even among some who profess themselves philosophers; and as they have been extensively evinced
in regard to geology and other branches, so they

have now influenced the discussion as to intelligent and moral inhabitants in the planets, as a question of *religious* import; and have led to the denial of their existence as derogatory to man's exclusive spiritual privileges and moral supremacy.

But the true spirit of the inductive philosophy tends to teach a modest and just estimate of man's place on one of the smaller planets in a subordinate solar system of a subordinate stellar cluster, his whole race occupying but a speck in space, and *as yet* a speck in time: while it points by analogy to other similar worlds, possibly in different phases of development, whether corresponding to *future* stages of the earth's progress, or to *past*. Inductive view.

I have elsewhere* endeavoured to illustrate and maintain the simple proposition, that *whatever is* animal in man's nature must be viewed as part of the same physical development and system, as the rest of animated nature: *Whatever is* superior to this belongs to a different order of conceptions, and cannot be affected by any physical considerations. On this view there is an obvious inconsistency in the Relation of man to the system of the world.

---

* See Essay I. § ii.

desire to connect ideas of the spiritual nature of man with the laws of the material world, or to imagine the belief founded on them endangered unless it claim so uncongenial an alliance.

Man in his animal nature part of a series.

Equally groundless is the anxiety sometimes evinced to disconnect, as far as possible, even the *physical* history of man from that of the rest of the material world — to find breaks in the continuity of the order of nature — to represent what is called " the human epoch " in the world's history as marked and separated by some great gap or interval from all preceding epochs — to isolate man among animated beings on the earth, and to isolate that earth itself as his abode, among other worlds. The dogmatic assertion that " there is no * transition from man to animals " is clearly untrue in regard to the physical nature they possess in common; and, in this respect, we need not go even to the lowest form of savage life to find but too close an approximation to the brute.

Origin of civilisation.

In the same spirit of viewing different parts of nature as *disconnected* from each other, and from

---

* See Essay on the Plurality, &c. pp. 81. 88. 164.

more comprehensive laws, much speculation has arisen on the history of civilisation, but on very insufficient data, because we *know* absolutely nothing of its earliest epoch, or of the element of *time* so essential to its satisfactory investigation.   Much objection is felt to supposing man's condition *progressive*, or linked in one chain from the lowest savage life up to the highest civilisation and advancement; though it is not easy to see on what philosophical or rational grounds.

But, on such subjects, our limited historical experience perhaps hardly yet furnishes us with sufficient data on which to prove any theory.   Within the historic period, civilisation advances only by the slowest and most insensible gradations, and is communicated from one race to another.   We have no right to assume that its advance was ever more rapid, but probably slower nearer its origin.   The difficulty of conceiving the transition from absolute brutality to high civilisation — from sounds little better than the inarticulate language of beasts, to highly artificial combinations, the index of mind and abstraction, — arises, at least in a great degree, from the utterly insufficient ideas commonly admitted in

Want of evidence as to early civilisation.

s

accordance with the received chronology, as to the length of time necessary for such advances.

If, according to some able inquirers self-civilisation of savage tribes be held to be impossible, and it be deemed necessary to refer to some higher source of enlightenment, I would merely remark (so far as the present subject is concerned) that this would not imply any interruption of natural order or the action of ordinary causes, being wholly confined to the province of the mental or moral world. Such improvements would of course be communicated through certain gifted minds, raised up for the purpose; and any development of the moral and intellectual nature of man, in proportion as it is traced to a higher source and considered to belong to an order of things distinct from anything material, it is the more clearly seen, can in no way affect or interrupt physical continuity.

Higher principle in man distinct from all physical considerations.

If it be affirmed that "man differs in his kind, and even in his order, from all other creatures," it is certainly not in his material nature or animal instincts, but only in a higher sense. If it be asserted that the introduction of reason and intelligence upon the earth "is no part nor consequence of the series of animal forms ; it is a fact of an entirely new kind : the transition from brute to man does not come within the analogy of the transition from brute to brute; "—this can only be understood as referring to man's higher nature, and not to that part of it related to physiological or material considerations : —if in man's intellectual and moral development there is a new principle super-added, this is a metaphysical conception apart from any physical conditions. But if such a principle be superadded in one race of beings or in one world, there can be no physical reason why it should not be so in other races and in other worlds.

To give a more precise illustration : — IF it were physiologically true that there were any peculiarity in man's *organs of utterance,* enabling him to frame articulate sounds, which is wanting in apes, then the cause of his superiority in this respect would clearly come under the dominion of *physical law,* and would mark a place and grade in *the connected scale of animal organisation.*

Illustration of this distinction.

Or, again, IF any peculiarity could be shown in man's *brain* to confer powers of *abstraction, moral consciousness,* or the like, which is deficient in the

animal brain, this in like manner would indicate a
clear physiological distinction, and would bring the
case under the category of *degree of physical organisation, or development.*

But, on the other hand, if no such anatomical distinctions exist, then the source of the difference
would be, as clearly, one *beyond the range of physical
science* or material analogies. And thus *either way,*
we must fully recognise the law of *continuity* as connecting man with the rest of the animated world:
in the one case, because the transition would be
simply one in *physiological* character; in the other,
because there would be no break of a *physical*
nature at all.

The relation of the animal man to the intellectual,
moral, and spiritual man, resembles that of a crystal
slumbering in its native quarry, to the same crystal
mounted in the polarising apparatus of the philosopher. The difference is not in physical nature, but
in investing that nature with a new and higher
application. Its continuity with the material world
remains the same, but a new relation is developed
in it, and it claims kindred with ethereal matter and
with celestial light.

Topics of this kind possess an interest chiefly with reference to the place they occupy in the wider question. The progress and development of the human race and of our little planet with all its accessories, form a part, and but a very subordinate part, in the process of evolution of the order of the universe. *In so far as* man's nature and capacities are *physical,* we may safely regard them as involved in the process of development of the physical world, without in the least endangering the dignity of that higher *moral* and *spiritual* progress, which, in proportion as it is held to be of an order distinct from the physical, must be admitted to be wholly independent of it, in its source, its cultivation, and its aspirations after perfection, and which can be in no degree affected or compromised by any speculations as to physical evolution, either in our globe or in the whole system of which it is so insignificant a member.

Again, — the essayist would make the earth in fact the boundary planet between those too near the sun, and those too remote, to be capable of intellectual or moral life on their surfaces. He would

Condition of the earth unconnected with spiritual nature of man.

represent it as alone of the proper density for mind
to grow upon — as enjoying the precise proportion
of light and heat for moral feeling to ripen, and the
exact degree of atmospheric pressure under which
spiritual aspiration can ascend!

Now on the *material* hypothesis there might obviously be a reason and a consistency in insisting
on these differences between a hotter or colder, a
moister or a drier planet, regarded as the parent
soil of mind and spirit. But this connexion is
entirely wanting when that hypothesis is so strenuously denied, and the essential discontinuity and
absence of all relation and dependence between the
development of man and the physical evolution of
the material world so strongly asserted.

If the highest aspirations of man, the relations of
his spiritual existence, be of a kind wholly independent of all physical evolution, and the very
conception of them derived from teaching of quite
another kind than any physical philosophy can
supply, it would then be a question wholly alien
and irrelevant, whether the earth were hot or cold,
moist or dry, solid or aërial; the greatest or the

least of planets—the sole inhabited world, or the most insignificant among the myriads of a peopled universe.

The question agitated in the publications before us has not been without its partisans among foreign writers; though it must be said in general that science on the Continent is happily kept far more strictly on its own ground, and free from theological bias, than among ourselves.

Œrsted (in the work before cited) alludes to some writers who, as he observes, "from one-sided religious or poetical views" have of late denied the existence of rational beings in other parts of the universe, in order to exalt the exclusive dignity of man.

Œrsted's views.

He glances at the different races of inferior beings who in past epochs have tenanted the earth, and infers by parity of reason that other races superior to man in his present condition, may at future periods in like manner arise. He then proceeds to argue very much on the same general kind of mixed metaphysical and religious ground as his opponents appear to have taken, that

" Our entire system * has developed itself in a
series of natural periods similar to the earth, and
that each planet must still submit to a succession of
creative transformations : consequently, we may infer that they have all had a succession of created
beings, with such variations only as the different
natural conditions of each must induce. Would it
not be a strange assertion that neither the older
planets at the most remote distance from the sun,
nor the younger and nearer ones, had any of them
attained to such a degree of development as is exhibited on our earth ?   Though a slight colour of
support might be given to the assertion, it could
never bear a close investigation. . . . Our system
is but a small part of a far higher system, with
which it has been developed under similar laws. . . .
And must we believe that on none of these planets'
similar or dissimilar to our own globe, reason has
been awakened to self-consciousness ? . . . . Thought
never finds repose, but rises to higher and higher
worlds ; and except on earth, can it recognise
nothing but barren solitude where no reasoning

---

* Soul in Nature, p. 53.

being has ever penetrated? No, it belongs rather to the nature of things that reason should develop itself into self-consciousness, not only in one spot but in every member of the system, although in different degrees. . . . . If we regard the whole of existence as a living revelation of Reason in time and space, we can conceive that the most varied degrees of development may be found distributed through all time; and that some bodies are still spheres of vapour, others have reached fluidity, while others have gained a solid nucleus, and so onward to the highest point of development, and then backwards again even to those bodies which are on the verge of final destruction. But even were it possible to maintain that self-conscious reason alone existed on earth, it still remains true, and is proved by the remains that have reached us of an earlier stage of development, that there was an immeasurably long period of time before the creation of man. Is it possible, then, that during the whole of this long period there was not a single being capable of perceiving and apprehending his own existence?" *

---

* Soul in Nature, p. 54.

I have given this passage at length, in order that
it may be seen how far speculations of a kind necessarily somewhat vague may influence a mind of such
philosophic capacity as that of Œrsted.   At the same
time, without professing to admit the entire force of
such reasoning, I am disposed to concur in the
general conclusion as at least a far more consistent
and worthy belief than that which would narrow
and restrict all intellectual and moral existence to
the confined limits of our little planet.

Final causes
appealed to
in this ar-
gument.   But apart from these speculations, other considerations of a more distinctly theological kind have
formed throughout the acknowledged basis of the
reasonings of both disputants, and indeed the main
motive for pursuing the inquiry.   In the first place,
the whole discussion has been closely connected
with the *argument from final causes.*

As this mode of argument is avowedly and extensively adopted by one writer, and, though much
restricted and qualified, is yet in some sense referred
to by the other, it will be desirable to recur briefly
to the general grounds on which such a line of argument can be sustained ; but as this is a topic which

has been already dwelt upon in a former essay *, it will not be necessary here to do more than refer to what has been there said, and to proceed upon those grounds to apply the argument to our present subject.

In the first place, the profound generalisations of Professor Owen have been referred to, who, in his discussion of the vertebrate skeleton and its theoretical archetype, dilates on the conclusion, that besides the organised structures *actually developed* on the plan indicated, according to the same principle the *rudiments* of an infinite variety of other such forms exist, and may therefore possibly remain to be developed; he observes that such conceivable forms are far from being exhausted in existing or past life on this globe; and that, " though they may never be developed as such in this planet, it is quite conceivable that certain of them may be so developed, *if* the vertebrate type should be that on which any of the inhabitants of other planets of our system are organised." †

Again, carrying out this idea to the structure of

Reference to Professor Owen's archetypal theory.

---

* See Essay I. § v.   † On Limbs, p. 83.

the eye and its possible modifications, as connected
with the vertebral theory, the author argues that in
Jupiter, with such provision for illumination as exists
there, and with the same laws of light and other
similar conditions, *analogy* would lead us to infer the
probability of beings with eyes conformably modified.;
and such creatures, he says, " *may* exist to profit
by such sources of light, and *must* exist IF the only
conceivable purpose of those beneficent arrangements is to be fulfilled."

Now, I quote these words more especially with
a view to remark *the nature of the reasoning :* we
cannot but observe the truly philosophical tone of
caution, united with the legitimate adoption of analogy, with which the distinguished author pursues
his conjecture. The reference to the archetype is
simply one of the highest forms of *conjecture* from
analogy, and supplies the same kind of antecedent
presumption, which the existence of a theoretical
mathematical formula would do in guiding us to a
physical truth.

That other modifications of the primeval type not
carried out into actual being on our planet, may
possibly be so in others, is abstractedly a very fair

conjecture; and the existence of such unrealised cases here may no doubt afford something like a presumption that they may be realised in other planets. So far, then, the reasoning is simply and strictly of an inductive character.

But the further remark, such beings " *must* exist," is, with equally just and philosophic caution, qualified by the condition, IF the purpose " *is to be* fulfilled." In a word, THE ARGUMENT FROM FINAL CAUSES is here kept properly distinct from that of INDUCTIVE ANALOGY, and is only maintained on the express *hypothesis* that we may reason at all from a purpose to be fulfilled.

On the general admissibility of such a reference to final causes, we must recur to the observations made in the former Essay. If the principles there laid down are admitted to be just, we shall the more readily acknowledge the general impropriety of attempting to solve a philosophical problem like that before us on any other grounds than those of legitimate physical analogy.

Improper introduction of final causes in science.

The argument from final causes is, in one sense, wholly distinct from any of a purely philosophical or positive kind; in another, it may be understood as a simple extension or higher theoretical view of the argument from analogy.

Thus, for example: Kepler argued from final causes to his first conjectures of the laws of the planetary motions. " I reasoned," he says, " that if God had adapted the motions to the orbits in some relation to the distances, it was probable that he had also arranged the distances themselves in relation to something else." But this was nothing more than a *guiding conjecture*, leading him to try *inductive* processes.

With more special reference to the question now before us, we find in numerous instances a purpose answered: we infer it *probable* that in others, or in all, it may be so likewise. In one case, we trace a structure adapted to a particular end; in another, under apparently analogous conditions, we infer that a similar end *may be* answered. In our earth there is a certain provision of light, and there are beings with eyes adapted to enjoy it. In Jupiter there is a certain, but different, provision of light: by analogy, there may be beings with eyes equally capable of enjoying it, though in a different degree

The conjecture is perfectly fair and philosophic

in its nature ; but other concurrent circumstances
must be taken into account before we can obtain
any higher amount of probable evidence. Nevertheless, as far as it goes, as *suggestive of presumptive
probability,* it is strictly legitimate.

But the subject has been carried out further by
the introduction of a still more metaphysical kind of
argument deduced from the " archetype " considered
as a revival of the Platonic idea of such archetypes
existing in the Divine mind.* On this point the
authority of the erudite Cudworth has been appealed to ; and the quotation of a passage from the
" Intellectual System " has given rise to some discussion, which, however, seems to me to have little
real bearing on the question. I merely observe
that the argument thus derived from our belief in
the Divine attributes or the assumption of inten-
tions or ideas in the Divine mind, whatever may be
thought of it in a metaphysical or theological sense,
does not, in my opinion, belong to the province of
physical philosophy; nor can it, I conceive, be
legitimately introduced in any such discussion as
the present.

---

* See Brewster, p. 84.   Supplement to Essay, p. 27.

The argument from final causes is largely appealed to by Sir D. Brewster, and is closely connected with the peculiar religious turn he is disposed to give to the question of inhabited worlds.   The belief that the planets and even the stars are inhabited, is upheld by him on the express ground that the probable end or purpose of their existence is no other than the support of organised life and of intellectual and moral creatures; and that to reject such a belief is to involve the irreligious idea of denying the final cause of their creation.

Thus, it is argued, that large globes, attended by an apparatus of satellites, or rings, *must* have been created for some great and worthy *purpose*, and that *we cannot conceive* any such purpose but that of sustaining animal and intellectual life.*   Again, as it is the obvious *function* of the sun to supply heat, so there is no *conceivable function* of the planets but that of supporting inhabitants.†

The fixed stars "were not planted in space to shed their light and heat upon nothing." ‡   "If the stars are not suns, for what conceivable purpose

---

* Brewster, p. 84.   † Ib. 90.   ‡ Ib. 159.

were they created?"*   Again, if no life existed
in the universe, he observes, the celestial movements
would be going on "fulfilling no purpose that
human reason can conceive—lamps lighting nothing,
fires heating nothing, waters quenching nothing,
breezes fanning nothing; and everything around,
mountain and valley, hill and dale, earth and ocean,
all *meaning nothing.*"†   Or, more pointedly, thus:
" In peopling such worlds with life and intelligence,
we *assign the cause of their existence.*"‡   "Life was
not made for matter, but matter for life; and in
whatever spot we see its atoms, whether at our feet,
or in the planets, or in the remotest star, we may
be sure life is there; life to enjoy the light and
heat of God's bounty, to study His works, to recognise His glory, and to bless His name."§

Now, in looking at the application of this kind
of argument in the present case, even if disposed
to admit the *truth* of the conclusion, I should still
have much doubt as to the *mode* of arriving at it.

Such an argument, in the first instance, necessarily *presupposes* the fact that the conditions of the

Objections
to this
reasoning.

---

* Brewster, p. 232.   † Ib. p. 181.   ‡ Ib. p. 179.
§ Ib. p. 180.

T

planetary bodies *are adapted* to be inhabited: the very point in question. But supposing the fact of such *adaptation* admitted, the next step in the argument, is the assertion, " that *no other* end or purpose of the existence of planets can be conceived." Now, this must, on the slightest consideration, be allowed to be at least a very hazardous assumption. How can we undertake to affirm, amid all the possibilities of things of which we confessedly know so little, that a thousand ends and purposes may not be answered, because we can trace none, or even imagine none, which seem to our short-sighted faculties to be answered in these particular arrangements ?

Supposing, however, that all this were conceded, it still remains to connect it with the conclusion because *no other end* can be assigned, therefore, *this one* end of sustaining life not only must be the *sole* real end which the Creator had in view, but must be actually accomplished in all the planetary worlds: an alternative and a conclusion which hardly appears warranted by any sound principles of reasoning.

The earth certainly was for myriads of ages

destitute of human beings : it existed, therefore, in *vain for man*. Hence, it is an undeniable parallel, if the planets are now uninhabited, and therefore useless, so was the earth for an unlimited number of ages in past epochs. If the one be a contradiction to final causes, and to be rejected as inconsistent with the Divine beneficence, so must the other be : yet this other we know to be the fact.

On the other hand, it is truly satisfactory to find so able a writer as the essayist joining his own testimony to that of Professor Owen against the narrow view of final causes, and beginning to avow that more truly just and philosophical principle of openly confessing that we know not, and ought not to pretend to know, *why* this or that arrangement is made. The essayist has, in fact, elaborately argued this point in his eleventh chapter, and more briefly and boldly in one of his dialogues * observes, " I do not pretend to know for what purpose the stars were made, any more than the flowers, or the crystalline gems, or other innumerable beautiful objects."

Opposite views more philoso- phical.

---

* Supplement, p. 5.

T 2

Again : " I have learnt much from Mr. Owen. I
have learnt from him, in many most striking cases,
to admire purpose in organic arrangements, when
purpose is apparent. But I have learnt from him,
also, that to infer facts from ' an only conceivable
purpose ' is a very hazardous process." *

Without any disparagement to Professor Owen,
I conceive, not merely this partial lesson, but a more
extended one as to the whole ground of argument,
might have been rather learnt in the school of
Bacon, and the incongruity of narrower views with
the essential spirit of induction : which would lead
us rather to recognise uniformity of plan, *law, order,
and unity, as the true exponents of design,* than to
seek for mere utility and ends to be answered, —
however important in a subordinate sense, — and to
carry out such principles as our only safe guides in
speculations even in the region of imagination,
whether exercised in peopling worlds or in de-
populating them.

Yet unphi-
losophical
conclusion

Yet, notwithstanding these admitted considerations, the argument of the " Essay," in fact, rests

---

* Supplement, p. 30.

much on final causes.   It is with reference to *an* on the other side.
end, to the fond belief in the high importance of *man*
in the universal scale of being, that so much stress
has been laid on his *recent* date upon earth, and the
alleged extraordinary peculiarity and singular isola-
tion of his position apart from all admixture with the
animal creation; and this is more especially dwelt
upon with a view to the argument that all the universe must be supposed a waste in order to enhance
the moral dignity of one puny race, and to enable
him to believe that his little world and his species
are the exclusively favoured objects of the Creator's
care.

The earth was for myriads of ages a void, and for
equally long periods tenanted only by inferior creatures, solely to the end that man might at length
come in solemn pomp at the close of the long proces-
sion, and take possession of his throne! and not only
so, but all the most distant planets and remotest
worlds, invisible to his eye or to his telescope, are
destined to a similar humiliating inferiority, solely to
swell the triumph of his supremacy, and to exalt the
dignity of that little speck on which the mighty

displays of Divine power and mercy were to be
made for his exclusive benefit!

Thus, then, we find, in point of fact, the argument from final causes applied with equal force to
support diametrically opposite conclusions. Tacitly
referred to on the one hand, it clearly evinces the
uninhabited condition of all the worlds but our own,
because man alone is privileged to be the exclusive
recipient of the Creator's beneficence ; openly and
strenuously upheld by the other disputant, it as
manifestly shows that the planets, and even the
members of the most remote sidereal system, must
all be teeming with rational and spiritual beings to
exalt the same Creator's perfections, and render a
reason for their existence.

On the one hand, it is argued that the planets
must be inhabited, because they could only have
been created for the sustenance of life; on the
other, that they *must be uninhabited,* because they
could only have been created as foils to enhance the
dignity of the earth and of man.   On the one side,
the universe must be inhabited because a void
universe would be *useless ;* on the other, a void

universe is *necessary* for the exaltation of man and of the Divine dispensations towards him.

If in a more wide and worthy sense we come to consider the religious application of the argument, whether for or against other inhabited worlds, it will be easily seen that, under any point of view, it amounts to little. With regard to the great truth of natural theology, the evidence which cosmical order affords for a Supreme Intelligence is in no way affected by the question of a plurality of worlds ; it stands unassailable on the basis of demonstration ; and can be little affected by any further speculative arguments. If the existence of inhabitants in the planets were as much demonstrated as on the earth, it would undoubtedly enhance the great argument by the extension of its evidence which would be furnished by the existence of organised structures, or of intellectual and spiritual beings, from one member of the system to many others : and if *universally* proved, it would tend to exalt this branch of the argument in a proportionate degree — by the infinite multiplication of such instances of physical, moral, and spiritual existence.

Bearing of the argument on natural theology

No real addition to the evidence; but wider contemplation opened.

But *demonstration* has never been in the slightest degree pretended to in this matter; the very utmost which its warmest supporters have claimed has never been more than *analogical probability*. Its force, then, as an argument of natural theology could weigh nothing in comparison with those substantial evidences which the *demonstrated* facts and laws of science afford. The utmost that can be said of such a theory is that, if admitted, it affords a beautiful opening for a more extended religious contemplation of the Divine beneficence reaching to so many more myriads of creatures capable of estimating it. Indeed, without unduly pressing the argument, we might fairly agree with Sir D. Brewster that, when we contemplate the combination of worlds upon worlds, and especially the movement of the entire solar system round the supposed *central sun*, " the mind rejects almost with indignation the ignoble sentiment that man is the only being that performs this immeasurable journey; " and that the planets with their train are but inert masses " mocking the creative Majesty of Heaven." *

* Brewster, p. 123.

Adopting the hypothesis of inhabited worlds, the devout believer in supreme and superintending Wisdom and Goodness, would doubtless find his adoration exalted and enlarged in proportion as he conceived a more enlarged sphere for its manifestation, and believed that myriads of other beings peopling other worlds, however different in nature from ourselves, were at the same time rejoicing in the light of the same beneficent Fountain of Good. Yet, he would recollect that all this is purely *hypothetical*, and stands on grounds quite distinct from the grand primary convictions of the unity and harmony pervading those worlds, and the consequent recognition of a Supreme Intelligence.

I have before adverted to the views broached by Œrsted on the general question of inhabited worlds. Without professing to assent to all his opinions, it must yet, I think, be allowed that there is much force in some of his representations. He argues much on the *intellectual* capacities of the supposed inhabitants of the planets; and more especially contends that *necessary* truth must be the same to them as to ourselves: though great differences might exist as to their perceptions of natural phe-

Œrsted's view of the subject.

nomena; and even in the former case, there may be great differences in the strength of the reasoning faculty.

To the same effect he continues a passage * before quoted, with respect to their moral development; and even goes further, and contends for a community of *moral* laws throughout the inhabited universe. Setting out from the theory of moral obligation on our own globe as arising necessarily out of the position and nature of man and his relation to his Creator, Œrsted argues that the same must hold good with the inhabitants of other worlds, making due allowance for the actual diversities in their conditions.

I will give one short extract: —

" Throughout the universe there are beings endowed with the faculty of understanding, that they may be able to catch some sparks of the Divine light; and God reveals Himself to those beings through the surrounding universe, and rouses their slumbering reason by that Reason which reigns throughout the sensible world; nay, He gives them a deeper insight

---

* Soul in Nature, p. 96.

into material existence the more their own minds are awakened; and thus they find themselves placed in a ceaseless stirring development, which, after having reached a certain point, removes them further and further from the idea that the foundation of being is that which is palpable, and which leads them to acknowledge and view themselves, their spirits and bodies, as parts of one eternal organism of Reason." *

Without professing to adopt or even to under- True unity of worlds. stand entirely the ideas thus eloquently expounded, I would yet willingly express an assent to the broad principle involved; and in such generalised conception (warranted as I think by the soundest inductive principles) it is that I would recognise THE TRUE UNITY OF WORLDS; not a narrow restricting of all development of mind and soul to one minute speck in the universe, which appears to me as unphilosophical as it is derogatory to the worthiest conceptions of the Supreme Mind, but an enlarged admission of all worlds as harmonising parts in one great whole, and of the universality of

---

* Soul in Nature, p. 109. See also p. 128.

such development of life, either " *in esse* " or " *in posse*," in actuality or in potentiality, as wide as is our belief in the universal presence and operation of the Great Source of all life, all mind, and all soul.

Including origin of worlds and past changes.

If we look back to past changes and the probable preparation for organised life even in planets not at present fitted for it; or if we ascend to the still grander question as to the probable order or law by which such beings may have come, or may yet come, into existence, though utterly unable to give any positive reply, yet we cannot fail to combine every reflection upon such a question with the great law of continuity, and beyond all doubt to regard as highly probable, some intimate connexion between the series of physical arrangements of unorganised matter, the successive gradations of organised existence, and the crowning of organisation with animal and intellectual life, making it the fit recipient for higher spiritual manifestations; and to recognise throughout the series the close dependence of the whole on some great principles of law and order (however unknown to us) to just the same extent as we acknowledge their dependence on supreme intelligence and power, exhibited to us through these its *cosmical* manifestations.

The precise nature and order of those causes which brought about the evolution of organised life on our globe are *as yet* unknown to us, however open to plausible conjecture. But in proportion as they might be known, or even rendered probable, they would afford *increasing* evidence of supreme intelligence : *increasing,* just as a more complex self-adjusting machinery would afford higher proof of intelligence than that which wants manual regulation. I have observed in a former place that all rational natural theology proceeds by tracing the steps and processes in which design is evinced. The more steps in such processes we can trace, the more satisfactory our convictions ; and if, where we do not *know,* we can fairly *conjecture,* the legitimate conjecture will have a like tendency.

Sir D. Brewster*, however, considers it a highly objectionable idea, to suppose the planetary system "*manufactured*" out of a nebulous mass by means of certain material laws. I, on the contrary, would

Secondary means the evidence of Divine operations.

Objections.

More Worlds than One, &c. p. 249.

accept this phrase, and contend that it is precisely the idea of their being *manufactured* instead of *made*, which would constitute the stronger proof of *intelligence* and *mind*.

Precisely in proportion as a fabric manufactured by machinery affords a higher proof of intellect than one produced by hand ; so a world evolved by a long train of orderly disposed physical causes is a higher proof of supreme intelligence than one in whose structure we could trace no indications of such progressive action.   And in proportion as we might be able to follow out more and more details of such a succession of causes, should we derive increasing evidence of that great truth.

Bearing of these views on revelation.

But the religious contemplations connected with this subject have assumed also more definite forms, and have involved difficulties, at first probably little to be suspected, on other grounds than those yet adverted to.   It has not been merely with a reflection on the enlarged beneficence of the Creator, that religious men have been contented to regard the supposed existence of planetary beings; they have also viewed the question in its bearing on the belief in a Divine revelation and the mysteries of human redemption. In this point of view, then, we must proceed to look at it.

In fact, the main object in view, in both the works under consideration, is in application of this theological nature, and to furnish replies, though in very different ways, to certain objections felt on religious grounds to the doctrine of a plurality of worlds.

Religious difficulties felt.

It has been held that the belief in the existence of rational and moral beings, however unlike ourselves, in other planets or other systems, is a notion which, apart from its physical vastness and difficulty, involves the believer in religion, whether natural or revealed, in perplexities and objections of the most serious nature, such as, in fact (it is alleged), seem only capable of being relieved by the rejection either of religious faith or of the idea of a plurality of worlds. These difficulties and objections are dwelt upon with great emphasis, and are stated at large by both writers.* We learn that they have pressed upon the serious convictions of many excellent persons, not only of high religious feeling, but even of cultivated

---

* Essay, §§ 2, 3. 4.   More Worlds than One, ch. vii. &c.

and enlightened minds; and have thus engaged the
attention of powerful champions of Christian truth
in the endeavour to remove or mitigate them.   In
particular, the arguments of Dr. Chalmers are referred to in detail; and while they are highly commended by the author of the Essay, are yet deemed
hardly sufficient without the further extension which
he conceives his own speculations confer on them.

Alleged
difficulties
in some in-
stances
vague and
unmeaning.

Thus, in a kind of ironical tone, the supposed advocate of a plurality of worlds is represented as
putting forth his theory in a somewhat dogmatical
manner; and, after making his various assertions as
to the nature of the remotest planetary systems as
the seats of animal and vegetable creations, and the
habitations of rational and responsible beings, is
made to add, " the only matter which perplexes us
is, that we do not quite see how *to put our theology
into its due place* and form in our system." *   Thus
far, however, we do not exactly make out what the
precise source of this perplexity may be, or how it is
that theology can want a local habitation in a planetary or sidereal system at all.

---

* Essay, p. 121.

We read through pages of eloquent and devout declamation, abounding with varied expressions of sentiments of religious humility and pious reliance on the care of the Creator—of profound conviction of His beneficent superintendence over all His works, and especially over man, so important in his *moral* and *spiritual* relations, even though insignificant in a physical point of view, and as measured by the immensity of the *material* creation—all which cannot but meet with cordial and general concurrence; but still in which there seems little to lead to any possible collision between such sentiments and any astronomical or cosmical speculations.

Again, the same "religious difficulties" seem to have made an equally powerful impression on some minds, whose perplexities are considered by Sir D. Brewster. A large portion of his work is devoted to the consideration of them; and it appears also that similar objections had presented themselves even in past times, when the speculations of Fontenelle and others were broached, and were of a nature to give a handle to the sceptics and scoffers of the day in their attacks on religion.

One of the first of these religious difficulties which is at all distinctly brought out, is the notion (which, however incredible it may appear, from the testimony quoted it would seem has really haunted some pious minds), that amid the multiplicity of creation, if believed to be replete with boundless worlds of life and intelligence, the humble individual, man, on this small speck of earth, *might be overlooked;* might be too small for the eye even of *Omniscience* to discern — too insignificant even for the vigilance of *infinite* and *omnipresent* Providence to take care of! — or that the *omnipotent* Creator " *may have too much to do* "* to extend his concern to all!

Every real religious doubt or scruple undoubtedly has a claim to be treated with the most unfeigned respect; but it is difficult to withhold an expression of astonishment when we find that such objections as these should have been seriously entertained, or have been deemed deserving of elaborate refutation by Dr. Chalmers, by an appeal (however forcible in itself) to the revelations of the microscope dis-

---

* See Brewster, p. 149.

closing worlds of animalcular life equally the objects
of providential care; or, further, that Sir D. Brewster should think this answer unsatisfactory, and
actually go through some amount of reasoning to
supply what he thinks a better!

But another difficulty next presses upon us of
deeper import; we are told that the assertion of
moral and spiritual beings in other planets is full of
danger to the belief in man's high privileges, the
possession of a special Divine revelation, the communication of spiritual gifts, or the promise of immor-
tality.   That these are, in fact, *exclusively* his portion
and inheritance, the security of which would be
hazarded by imagining any other claimants on such
dispensations of the Divine mercy. *Difficulty as to other worlds partaking in spiritual privileges.*

There are doubtless peculiar charms in the *exclusive* nature of privileges, which, in fact, often
constitutes the main value of their possession.   Thus,
in the present instance, a feeling of complacency
and comfort seems to be inspired by the belief
which, with a non-peopled universe around him,
man can securely entertain, that he is the sole
favoured child of his Creator, and can console him-
self in looking round on the untenanted planets with *Spirit of ex-clusiveness.*

the happy reflection, "all these vast worlds were made, if not indeed in a direct way for my use, yet still for gratifying me with the delightful reflection that my own race on my own happily constituted planet are alone permitted to enjoy the blessings of life and intelligence, which are denied to other inferior globes, or to aspire to any of those higher gifts of grace or glory which are our peculiar inheritance." We cannot fail to reflect how highly and peculiarly spiritual is such a contemplation, and how eminently worthy to be dilated on by a Christian divine! how powerfully tending to elevate at once the conception of a beneficent Creator and the moral dignity of the creature!

Moral training in other worlds.

Again, it is urged as a more specific objection, that, if we people the planets, we must by analogy suppose those inhabitants to have had a similar intellectual and moral progress and training to that which the human species have undergone. But this, it is alleged, would impugn the special character of God's government of our world, which consists only with "man's nature and place being unique, and incapable of repetition in the scheme of the universe."*

---

* Essay, pp. 36. 51.

The "moral training" and the "religious edu-
cation" of the human race and the like phrases,
have become very common with writers of late years.
I can only apprehend their meaning in the simple
sense, that, as mankind have advanced in civilisation and intelligence, different codes of morality
have been tolerated, and various forms and dispensations of religion suited to those successive stages of
advance have been established.   But on what
grounds it can be asserted that such a series of progressive movements are "incapable of repetition,"
if the circumstances of moral and spiritual beings
should require it, whether on our earth or elsewhere, after all that has been urged, I am wholly
unable to see.

To the same purport the author of the Essay
again observes : —

"Religion seems, at first sight at least, to represent man's history and position as unique.   Astronomy, some think, suggests the contrary: I examine
the force of this latter suggestion, and it seems to
me to amount to little or nothing."*

"Unique
position" of
man.

---

* Supplement, p. 42.

In what way religion "seems" thus to point to
the unique position of man, is precisely the point
which I fail to apprehend. I see nothing in natural
religion to lead to this conclusion ; and in the pages
of revelation, even in its most literal acceptation,
whatever privileges are conceded to man are surely
nowhere denied to other *possible* races of beings.

Assertion of
revelation
only made
to man.

But, more precisely, the representation takes this
turn; man, it is contended, is the peculiarly
favoured creature of Heaven—the earth in which
he dwells the scene of the most peculiar and transcendent displays of Divine mercy : the human race
has peculiar and exclusive relations with the Deity.
The history of this race attests a continued scheme
of exclusive dispensations of grace. And we are
told " The earth, selected as the theatre of such a
scheme of teaching and redemption cannot in the
eyes of any one who accepts the Christian faith, be
regarded as being on a level with other domiciles.
It is the stage of the great drama of God's mercy
and man's salvation ; the sanctuary of the universe ;
the Holy Land of the creation ; the royal abode, for
a time at least, of the Eternal King."*

---

* Essay, p. 44.

That such privileges are granted to man, that such displays of Divine mercy have been manifested, is no doubt clear to every believer; but why they must be supposed *exclusive;* why *limited* to man and his earth; why *denied to all other* possible or conceivable races of beings — I am at a loss to understand, or to find a particle of evidence in support of so extraordinary an assertion.

It is, professedly, to meet the serious difficulty which is thus believed to result, if we deny the *exclusiveness* of human privileges, that the essayist has engaged in the arduous task of attempting to prove that there *cannot* exist in any of the celestial bodies inhabitants of a moral or religious nature to dispute with man his *exclusive* privileges. If this conclusion be made out, of course the objection, such as it may be, falls to the ground. But if the point of uninhabited worlds be not *demonstrated* (and who can pretend to say that it is?), then it would be implied that the whole faith of the Christian world is in jeopardy: and its vindication and truth are staked upon the acceptance of the belief in an uninhabited universe, which can be nothing more, at best, than a visionary speculation!

Constitutes no real objection.

But if we look further at the real nature of this difficulty, and endeavour to put it into more definite shape, I conceive it can only be stated somewhat in this way : — A special manifestation of the Deity in the Gospel dispensation is affirmed to have been vouchsafed to the inhabitants of the earth. But the earth is a very small and insignificant unit in a vast universe of similar and greater worlds, all teeming with unnumbered moral and spiritual beings possibly of far higher dignity than man. *Therefore,* we are to doubt the reality of the revelation to us!

Antecedent probability the same for other revelations.

But, perhaps, it may be said the objection only refers to the *general antecedent probability* of a revelation (in the accepted sense of the terms), and does not descend to the question of its particular alleged *evidences.* Now, as to this question, we have only to consider in what possible way the assertion or denial of inhabitants in other worlds can affect the antecedent probability of a revelation being made to the inhabitants of the earth. Taking the argument for probability as stated by the most approved writers on the " evidences," it can be maintained only on the *same general grounds* of the Divine attributes, which would render it *equally admissible* in regard to the supposed inhabitants of any other worlds — the creatures of the same Supreme Power, to whom, by parity of reason, it must be supposed a revelation *would* be equally granted by the same Supreme Goodness *if* they needed it, and which assuredly *could* be accomplished under whatever diversity of condition by the same Universal Omnipotence.   On the other hand, if those grounds of argument, so commonly adopted, are themselves thought unsatisfactory in regard to the planets, why are they admissible with regard to the earth?

If analogy did assign intelligent and spiritual inhabitants to any number of other worlds, or if it represented the whole universe as nothing but one teeming creation of moral and spiritual existence, surely the same analogy would not only admit, but rather require, the extension of the same argument for the probability of Divine communication with one portion or race of such beings as with another; or rather, it might even seem to suggest the notion of one grand universal manifestation of the Divinity in all the worlds over which the same universal Providence presides.

Analogy would extend revelation to other worlds equally

But the objection has taken a more specific form

Objections

from pecu-
liar views
of man's
condition.
from a reference to certain deep doctrinal views relative to the moral and spiritual condition of man,
and especially the sinfulness of his nature, which, it
is alleged, are endangered by the consideration of
other inhabited worlds. *Yet surely*, if those beings
are in their spiritual needs similarly circumstanced
to man, it is the fairest presumption that corresponding means of spiritual restoration would be
granted them. If they do not require those means,
they may equally stand in other relations adapted to
their nature and condition. Yet this kind of objection seemed to Dr. Chalmers so formidable as to
require elaborate refutation, and Sir D. Brewster
also makes a lengthened reply; in the course of
which he plunges into the depths of a metaphysical
theology, the theory of original sin, and the hopeless
question of the existence of evil; which last he
seems disposed (unless I misunderstand him) to
solve in the convenient way so readily adopted by
some other philosophers and divines, of *denying its
reality*, and affirming that "the spectre of moral evil
has been conjured up by ourselves." *

---

* Brewster, p. 138.

Without pretending to go into such discussions, which appear to me equally needless and interminable, or dwelling on the hazardous nature of this mode of reply, it will suffice to remark generally here that these topics seem chiefly connected with the present subject and with the question of the origin of the human race only from the prevalent adoption of the theology of a peculiar school, according to whose system certain supposed *physical* changes were induced by sin and "the fall;" to which in vulgar estimation such singular effects have been attributed. It ought surely to suffice a reasonable and Christian inquirer to refer to the language of the Apostle Paul *, and to perceive that such ideas can find little foundation even in the most literal acceptance of his words, which do not contain the smallest allusion to any *physical changes in man's nature,* but to a subjection to death in Adam *in the same sense* as accords with a deliverance from it in Christ.

Origin of much of the supposed difficulty in peculiar views of Christianity.

In pursuing the argument, however, Sir D. Brewster puts the difficulty, at least more distinctly

More precise statement of the main difficulty.

* As *e. g.* Rom. v. 18, 19.

than before, into the mouth of a supposed timid
Christian, " How can we believe that there are inhabitants in the planets when God had but one Son
whom He could send to save them ? " and adds, " If
we can give a satisfactory answer to this question,
it may destroy the objections of the infidel, while
it relieves the Christian from his anxieties."   To this
task he therefore addresses himself, putting the case
thus : — " When at the commencement of our era
the Great Sacrifice was made at Jerusalem, it was
by the crucifixion of a man, or an angel, or a God.
If our faith be that of the Arian or the Socinian,
the sceptical and the religious difficulty is at once
removed ; a man or an angel may be again provided
as a ransom for the inhabitants of the planets.   But if
we believe with the Christian Church, that the Son
of God was required for the expiation of sin, the
difficulty presents itself in its most formidable
shape."*

Sir D.
Brewster's
answer.

The author's answer is to the effect that, " as by
some process of mercy, which we understand not,". .
the saving power of the sacrifice has been communi-

* Page 139.

cated alike to the most distant nations and ages, past and future; so it might just as easily be communicated to the inhabitants of the most distant planets and worlds; and if this should not be convincing, another answer is hinted at (though not entirely approved by the author, yet) as satisfactory to some minds, *viz.*, that the Divinity might in other planets " resume a physical form, and expiate the guilt of unnumbered worlds." *

These suggestions may be safely left without further comment. Minds so constituted as to feel such a difficulty will probably be well satisfied with the solutions here proposed.

Source of difficulties in narrow views of Christianity.

Others, however, have expressed a similar difficulty more briefly and emphatically, by observing that " the question is not merely one of a *revelation*, but of an *immolation of God* ; " and that " to imagine such an event *repeated* is an idea too monstrous to be conceived."†

But I would ask, taken literally, can such an idea *be conceived at all* by the human faculties, even in one instance? Can we in any sense *reason upon it* beyond the mere words of the sacred writers through

---

* Page 142.
† I quote these expressions from a letter addressed to me by a friend.

which alone it is disclosed to us so as to find any real contradiction in a supposed recurrence of such an event ?   Can we presume to say that such a *repetition* is of necessity implied, or to determine the case at all for other worlds, who after all might not need redemption ?

The same difficulties urged by John Wesley.

The same religious difficulties adverted to by both the writers in this discussion were urged with some force long ago by John Wesley*, who (like the

---

* I am indebted to a friend for the following extract from a sermon of John Wesley, Text, "What is Man?" Psalm viii. 3, 4.

"Let us then fear no more! Let us doubt no more! He that spared not His own Son, but delivered Him up for us all, shall He not with Him freely give us all things?

" ' Nay,' says the philosopher, 'if God so loved the world, did He not love a thousand other worlds, as well as He did this? It is now allowed that there are thousands, if not millions of worlds, besides this in which we live. And can any reasonable man believe, that the Creator of all these—many of which are probably as large, yea, far larger than ours—should show such astonishingly greater regard to one than to all the rest?' I answer, Suppose there were millions of worlds, yet God may see in the abyss of His infinite wisdom reasons which do not appear to us why He saw good to show this mercy to ours in preference to thousands or millions of other worlds. I speak this even upon the common supposition of the plurality of worlds—a very favourite notion with all those who deny the Christian Revelation ; and for this reason, because it affords them a foundation for so plausible an objection to it. But the more I consider that supposition, the more I doubt of it : insomuch that, if it were allowed by all the philosophers in Europe, still I could not allow it without stronger proof than any I have met with yet."

He then cites the argument of Huyghens, to the effect of the probability of planetary inhabitants, but proceeds to mention that at a later period Huyghens entertained doubts on the subject from observing that the moon had no atmosphere. Hence, he says, the argument falls to the ground. He then goes on : "But, you will say, suppose this argument fails, we may infer the same conclusion—the plurality of worlds—from the unbounded wisdom, and power, and goodness of the Creator. It was full as easy to Him to make thousands of worlds, as one. Can any one, then, believe that He would exert all His power and wisdom in creating a single world? What proportion is there between this speck of creation and the Great God that filleth heaven and earth, while

author of " The Plurality ") could find no other solution than by denying the existence of other inhabited
worlds, as an unfounded idea, merely taken up (as
he says) by infidel philosophers, as affording a plausible objection against Christianity.

But the whole objection turns on the same tacit
but unfounded assumption, that because the privileges
of redemption *are granted* to the inhabitants of this
earth, they are therefore *not granted* to those of any
other worlds : and that *it is* a part of Christianity to
hold this exclusive view.

The objec-
tion de-
pends on an
unfounded
assumption.

That such high privileges (as already remarked)
are asserted with respect to man on this earth in the
New Testament is manifest ; but it may be asked in
vain what particle of proof can be alleged for *denying
it,* with respect to *the possible inhabitants of other
worlds ?* Let those who urge this objection produce

" ' We know, the power of His almighty hand   ,
Could form another world from every sand ? ' "

" To this boasted proof, this *argumentum palmarium* of the learned infidels, I answer, Do you expect to find any proportion between finite and
infinite? Suppose God had created a thousand times more worlds than
there are grains of sand in the universe, what proportion would all these
together bear to the infinite Creator! Still, in comparison of Him, they
would be, not a thousand times, but infinitely, less than a mite compared
to the universe. Have done, then, with this childish prattle about the
proportion of creatures to their Creator, and leave it to the all-wise God
to create what and when He pleases. For who besides Himself hath
known the mind of the Lord, or who hath been His counsellor?"

No proof
of the ex-
clusiveness
of privi-
leges.

a single argument from the reason of the case (if they think it one amenable to reason at all), or a single passage from the New Testament in which *such limitation or exclusion* is asserted, and the objection may have some weight. But this they do not attempt, nor do they seem to perceive that the attempt is necessary to establish their case.

Inconsis-
tencies of
Wesley's
argument.

That Wesley should set down all philosophers who advocate a plurality of worlds as " learned infidels " is not surprising; but with his acknowledged acute powers of reasoning, it is remarkable that in earnestly asserting (what is not contested) that such high privileges *are* vouchsafed to man, and expatiating on their greatness and value, he seems to think it sufficiently proved that they cannot be granted to other races of beings.

And when he so strangely glances at, but passes over, the undeniable argument that *Omnipotence* could as easily *create* thousands of worlds as one, he seems equally blind to the obvious answer that the same Omnipotence could as easily *redeem* thousands of worlds as one, if they needed it.

Contradic-
tion in rea-
soning on

The truth is—all these and the like difficulties arise wholly from the inconsistency of *attempting to*

reason at all on subjects which the writers themselves at the same moment pronounce *to be above all* such a sub-ject.
reason : attempting to argue that the Deity could,
or could not, act in this or that way, when they, in
the same breath, assert the incomprehensibility of
His counsels.

If it be an inscrutable mystery *wholly beyond
human comprehension* that God should send His Son
to redeem this world, it cannot be a *more* inscrutable
mystery or *farther beyond* human comprehension that
He should send His Son to redeem ten thousand
other worlds. If, on the other hand, the *mystery*
be amenable to any reasoning, or charge of inconsistency or incompatibility with our conceptions in
the one instance, it must equally be so in the other.

Cases of this kind, it seems to me, can only be
viewed under one alternative. Either they are in-
effable mysteries of the spiritual world not to be
inquired into or reasoned upon ; or, they are modes
of expression adopted by the sacred writers fairly
amenable to rational criticism : in the one case, no
plea of difficulty or inconsistency is admissible ; in
the other, none can arise, or is easily explained.
And the declarations of Scripture express nothing

x

with respect to the case of other races or other worlds; addressed to our own race, they profess to declare only what concerns that race.

But the whole discussion cannot but suggest a passing remark on the style and tone of theologising evinced in the very statement, whether of the objections or the answers. They seem to belong altogether to a somewhat obsolete school, and to refer too much to those narrow humanised ideas of the Divinity and His dealings with man derived so commonly from too literal an interpretation of the anthropomorphisms of the Hebrew Scriptures, and little consonant with the more enlightened views which a better dispensation encourages.

Other instances of narrow views.

Though referring to a different part of the subject before discussed, yet I may just cite another instance of expressions evincing a similar narrow spirit, and seeming to imply ideas which might have been supposed exploded from the minds of philosophers : — " Can we believe that he who formed the *worlds* has made only *one*, and that in place of resting on the seventh day, He rested during the whole week of creation, and still rests, having transferred His almighty power to certain laws of matter and motion, by which the sun and all his planets were *manufactured* from nebulous matter ? " *

The common occurrence of language of this sort (not solely in reference to the present question) leads us to reflect generally how much of the unhappy perplexities and objections which beset the minds of believers on some points, especially where science is supposed to come into collision with religion, must be traced to the influence of popular dogmas, founded on a narrow literalism, which, as in the cases already glanced at, betrays its Judaical origin, rather than any connexion with the enlightening influences of Christianity.

There is one other idea of an extremely peculiar kind taken up by Sir D. Brewster†, referring to the question *where* believers can place the *locality* of their future state. He refers to some passages of Scripture, which literally seem to imply, or at least to countenance, the idea of a plurality of inhabited, or at least habitable worlds. He enters on a calculation to show that, in a future state, for the myriads of resuscitated

Peculiar views as to future state.

---

* Brewster, p. 249.   † Ib. pp. 18. 256.

human bodies, the earth would afford utterly insufficient room; and that the future abode of
man must, therefore, be in some of the other bodies
of the solar system, which *being habitable,* will be
suitable to this purpose. Such is the idea indicated
even in the introduction of his volume, and such
the final conclusion to which the whole discussion
leads.

It is the danger threatened to *this* doctrine which
constitutes the main cause of alarm at the triumph
of scepticism in the denial of a habitable condition
to the planets. Yet it might rather seem that their
being uninhabited would be more favourable to this
doctrine, as affording more ample space for the
reception of resuscitated humanity.

Difficulty
obviated.
    Though unable to perceive the importance or
reasonableness of this question, I am yet anxious
to give it the most respectful consideration; and
therefore feel bound to add, if it be needed, for
the confirmation of any wavering mind, that, in
my opinion, the slightest attention to the writings
of the Apostles affords a more satisfactory solution
of the difficulty than any astronomical theories
whatever. If there be one point clear in their

declarations, it is that the resuscitated body is represented as something at least wholly different from our present *material* nature.   It is sown a *natural* body — it is raised a *spiritual* body.   (σῶμα ψυχικὸν, ... πνευματικόν.)*

Such then, on the whole, is the formidable difficulty of a plurality of inhabited worlds!   Such the dispute which threatens the alternative of a surrender either of faith or of science!   Such the source of so much perplexity to thoughtful and religious minds, to the solution of which such elaborate speculations must be devoted!   Such the danger impending on Christianity, which it is the aim of the essayist, by such laboured reasonings and startling paradoxes, to avert!   and to escape from which Sir D. Brewster, by so opposite a route, would guide his readers!

General remarks on the application of the arguments.

It is difficult, perhaps impossible, fairly to judge of the *convictions of others,* and I would wish to treat all serious convictions with unfeigned respect; but after what has been already observed, I must confess myself more disposed to concur in the

---

* 1 Cor. xv. 44.

x 3

abstract *justice,* than to perceive the *consistency,* of Sir D. Brewster's remark, " The difficulties we have been considering, in so far as they are of a religious character, have been *very unwisely* introduced into the question of a plurality of worlds!"*

Recommendation of clearer views of the subject in its first principles,

For my own part I would rather disclaim the entire principle of such discussion, whether in the objections or the replies. The strange juxta-position of ideas of such very different kinds, religious and physical, which characterise these reasonings altogether, seems in a peculiar degree likely to expose the whole subject to the attacks of the scoffer. The expression of theological dogmas contrasts singularly when mixed up with the speculations of science. The languages of the two sound strangely together; and I am powerfully reminded of the wisdom of Bacon's suggestion, " DA FIDEI QUÆ FIDEI SUNT."* And if these speculations on

both in philosophy and theology.

planetary worlds have really caused any perplexity to persons capable of profiting by rational and philosophical views of the subject, I would rather

---

* Brewster, p. 154.   † De Augm. lib. iii. c. 2.

entreat their serious attention to the question, whether the cause of truth would not be better served by a preliminary endeavour to acquire clearer views of the grounds on which scientific speculation on the one hand, and the expression of theological doctrine on the other, are legitimately established and of the very different basis on which they respectively stand, each firmly on its *own* ground, but as unstable on the other's as a ship on land or a pyramid on the sea.*

To this end, then, — to aid in such an inquiry, — a very few concluding remarks may be directed.

The case as put by the essayist, " How to place our theology," with reference to the question of inhabited worlds, may perhaps be taken as in some sense the expression of a difficulty felt more widely as to the *relation of Christianity to physical science generally;* and is probably similar to that acknowledged by another eminent writer, who declares it to be " the great problem of the age to

Some general suggestions on this subject.

---

* That it is not unseasonable or needless to press the consideration of the *proper ground* on which such inquiries ought to be conducted, nor to insist on the independence of matters of faith from those of science, is further evinced by the announcement, while these sheets are going through the press, of another anonymous publication on the theological argument, " The Plurality of Worlds.—*The Positive Argument from Scripture,*" &c. London, 1855.

reconcile faith with knowledge, philosophy with religion." *

To discuss such a question in the way which its importance demands, would be to open a very wide inquiry, wholly beyond the necessary limits and scope of this Essay, and one which, perhaps, could hardly be entered upon with much prospect of satisfying the varieties of apprehension (or possibly misapprehension) which it might encounter. All I shall attempt, therefore, will be to offer a few general and somewhat fragmentary remarks, which must be left to the judgment of the reader to apply.

Plurality of worlds only a subordinate point.

In fact, the question now discussed is only a subordinate branch of a far wider subject. The mere inquiry as to the probability, or otherwise, of inhabitants in the planets, is in itself of a very secondary and unimportant character. What we have to consider is rather the broad principle involved in *any or all cosmical speculations*; they all tend to place our earth, and man upon it, in a very subordinate position in the vast universe; not merely in space and position, but in general re-

Subordinate position of man in the universe.

---

* Archdeacon Hare, Life of Sterling, p. 121.

lation; as one small and insignificant link in the vast chain or rather universally connected ramification of physical causes, of which there is no one constituent part more the *head* or the termination than any other. Hence a difficulty is felt by some because they imagine that moral and spiritual relations must follow the same law.

Yet nothing can be in itself more unfounded; and still more must this appear when considered in connexion with what has been here advanced. The tenor of the whole preceding argument has been to point out the *independence of the physical order of things, and the spiritual.* * It has been maintained that the very idea of a spiritual nature in man, *in so far as* it is independent of the body, belongs to a higher order of conceptions, of a kind radically different from, and forming no part of a physical system; and beyond all scientific reasoning.

It is to this class of conceptions that religious doctrines properly refer; and thus it would seem, *on general grounds at least,* unreasonable to expect that the two should have any connexion; or

Spiritual truths distinct from physical.

---

* See above, pp. 74. 242.

to be anxious, on the one hand, to frame theories combining physical science with religious belief, or, on the other, to imagine the region of physical truth an unsafe locality for a theological creed. But *in detail*, perhaps it may be said this will be found otherwise.

Connexion of science and religion in natural theology. There is, certainly, one point in which physical science and theology are obviously and unquestionably in close contact and dependence,—in the primary inference of a Supreme Intelligence as derived from the order of nature so largely dwelt upon in the preceding Essay*; and which forms the substantial and necessary foundation of all *rational* conceptions of religious belief. But this *foundation*, however solidly laid, rises no higher than to the lowest basement; and if the conclusions at which *reason* arrives, are restricted according to the nature of the physical evidence, while they may afford some corrective of too blind and literal a dogmatism, they offer no disparagement to higher spiritual convictions. In this respect, then, and to this extent at least, we may, perhaps, see " how to

* See Essay I. § v.

place our theology" on the basis of these sublime
deductions from physical science. These deductions
(as before mentioned) are confessedly *very limited
and imperfect*. They present nothing, as it were, Natural
theology
but the meagre skeleton; to fill it out with sub- very
limited.
stance and life is the function of those higher inti-
mations derived from moral and spiritual sources,
and which in their essential nature stand apart from
all physical considerations.

A scientific natural theology does not rise to the Natural
theology
aspirations of a spiritual or moral Theism; still less prepares for
revelation.
to the scriptural or ecclesiastical doctrines. It tells
very little of any Divine *attributes,* and NOTHING of
the *mode* of the Divine existence; but for that very
reason it presents *nothing to contravene* higher
spiritual views on these points when *proposed from*
OTHER SOURCES.

Yet, when, from this primary position, we advance Physical
difficulties
to those more precise views of religious doctrine, in more
particular
we find a disposition continually evinced to place religious
views.
them in connexion or in collision (as the case may
be) with physical considerations; to raise philosophical theories on a theological basis, or to find
fatal difficulties in the failure of such attempts.

Sources of misconception on these subjects. If we would trace such tendencies to their source, we may find a too common origin of misconception in the ignorance in which many, even of considerable scientific attainments, remain as to the real nature of Christianity; when their profession of it consists

Ignorance of Christianity. either in merely bowing to conventional requisitions, or is based on notions derived solely from the prevalent or established belief, instead of an enlightened and independent examination of it for themselves.

Want of sound philosophical principles. Again, many who are extensively versed in the details of a *particular branch* of science, may often have reflected little on its wider relations and philosophical spirit.   Hence, while they admit the impropriety of some of the speculations just adverted to, they are deficient in the distinct conceptions of the broad principles and grounds of all philosophical inquiry.

Applications of physical philosophy. Physical philosophy has doubtless within itself the *germs* of higher knowledge, and presents us with those first elementary notions which are pre-eminently valuable, as subservient to the establishment of theological truth on a rational basis.   In such a sense, and within very circumscribed limits, *theology deduced from philosophy,* may be sound and valid.

But in every case *philosophy deduced from theology must* be *essentially* erroneous and fallacious; it is no longer *philosophy.* It appeals to other authority, and disowns its proper inductive character.

The desire whether for peopling or for dispeopling planetary or sidereal worlds *on theological grounds,* appears to arise from the same fundamental misconception or disregard of the proper provinces and limits of philosophy and of theology which has led, in so many other cases, to an unhappy and incongruous mixture of the two, — producing nothing, as Bacon has so justly observed, but " a fantastical and superstitious philosophy and a heretical religion."* Of this mode of procedure we have had abundant instances in all stages of scientific advance.

Erroneous systems from misconception of first principle.

Without recurring to more ignorant ages, and the speculations of the schoolmen, we trace the very same spirit in later times in the formation of such systems as that of Tycho, founded on the idea of reconciling astronomy and Scripture; in the vortices of Descartes, deduced by reasoning on *theological* grounds from the perfections of the Deity; in the cosmical

Theological philosophy.

* De Augm. l. iii. c. 2.

theories of the Hutchinsonians, or what they termed
" Moses' Principia," founded on the Hebrew Scriptures, in opposition to Newton's; and in our own
Bible geo-logy. times in the various schemes of the Bible geologists, each in succession presenting but some new
shade or modification of the *same radical misconception* to take the place of its exploded predecessor.

It is worth while to dwell on this last instance as
very instructive in its consequences, especially to
those who have not antecedently taken more general
views. Even at the present day there are not
wanting occasional attempts to keep up the hopeless
chimera of erecting theories of geology on the
Mosaic narrative. It is needless to observe that, as
all notion of an accommodation of the *facts* to the
text has long since been given up by all *sane* inquirers, these attempts are now merely directed
to explaining away the sense of the text; in which,
they no doubt succeed by *such* principles of verbal
interpretation as, if fairly applied to other parts,
would readily enable us to put on any given passage
any required construction.

Contradic-tion be-tween geo- All inquirers, possessing at once a sound knowledge of geology, and capable of perceiving the

undeniable sense of a plain circumstantial narrative, now acknowledge that the whole tenor of geology is in entire contradiction to the cosmogony delivered from Sinai ; a contradiction which no philological refinements can remove or diminish ; a case which no *detailed* interpretations can meet, and which can only be dealt with as a whole.*

I have elsewhere† fully discussed this subject, and have there explained the only view which I think the case admits; in one word, that the narrative, as a whole, cannot be received as *historical,* but was a representation accordant with the apprehensions of the Israelites introduced as a basis for the institution of the Sabbath; while Christianity, I contend, can be in no way affected by any such contradiction to the Old Testament law ‡, with which it has been

---

* For some excellent remarks bearing on this point, the reader is referred to Mr. Kenrick's "Essay on Primeval History." London, 1846. Preface, p. xiv.

† See Connexion of Natural and Divine Truth, p. 245. While this edition has been in the press, I have seen a new discussion of the "six days" carried on with much erudition and more warmth, between Professors Tayler Lewis and J. Dana of the United States, on either side characterised mainly by the same *fundamental misconception* and confusion of thought in regarding the Mosaic narrative as if it were a real scientific theory the terms of which must somehow be tortured into accordance with physical facts.

‡ For support of this view see my Essay on the Law and the Gospel, "Journal of Sacred Literature," April, 1848, and Art. "Creation," Kitto's "Cyclop. of Biblical Literature," p. 485.

erroneously mixed up; on the contrary, the palpable discrepancy is valuable, as reminding us the more forcibly of its independence.

Encroach-
ment of
science on
erroneous
belief. It is undeniable that the advance of physical knowledge has from time to time made inroads on the territories which prescriptive error had once consecrated to religion. So the Copernican heresy not only deposed the earth from its proud immobility as the centre of the universe and the throne of spiritual infallibility, but set at nought the *letter* of numerous scriptural texts: it entailed the impious doctrine of *antipodes*, and destroyed the ideas of *an upwards* and *a downwards*, a *local* heaven *above*, and a *local* hell *beneath* the earth. It broke through the solid firmament, and placed in jeopardy the existence of a physical purgatory. Yet real Christianity has been in no way injured, but the reverse; its rational and spiritual character has been the more powerfully asserted and vindicated.

In all ages bigotry has erected its strongholds on the basis of ignorance, and especially on erroneous physical ideas; and its advocates have then resisted all advance of intelligence on the plea that it is destructive to the security of religion, as it doubtless is to their baneful systems; while they thus by clear implication dissever the claims of religion from those of truth, and degrade the profession of it to the level of the most baseless superstition.

Even among professed Protestants, and in a phi- *Favourable to a purer religion.* losophic age, men have hardly become convinced that the advance of physical enlightenment, so far from being hostile to religious truth, is eminently serviceable to it, were it only in dissevering it from false allies and equivocal auxiliaries, and thus exhibiting its true spiritual power, when cleared from the heterogeneous incumbrances and corruptions which a false philosophy or a narrow literalism had fastened upon it, and which, instead of aids and defences, are in reality its hinderances and disfigurements.

But when we turn to the pages of the Bible, it *Physical difficulties in the Bible.* is doubtless the fact that continual reference is made to *physical considerations* of various kinds, which may in many instances give rise to difficulties. These will be so very differently estimated in magnitude and significance by different minds, that it would be impossible to discuss the question in a way satisfactory to all. It may perhaps suffice to suggest

Y

a general application of what was above observed in
reference to the instance of the geological discrepancy.

General
rule for
physical
difficulties
in Scrip-
ture.
Whenever the sacred writers introduce *physical*
statements, they may fairly be understood as speaking *conformably to the existing state of knowledge,* or
adapting themselves to the ideas, belief, and capacities of those they addressed. In any such cases it
would be irrational for us at the present day to insist
too literally on such representations, and especially
to reason on them in cases where we are precluded
from examining into all the circumstances, or can-
vassing the evidence. But if in any instance the
letter of the narrative or form of expression
may be found *irreconcilably at variance with physical
truth,* we may allow to those who prefer it, the alternative of understanding them either as religious
truths represented under sensible images, or as a description of events according to the preconceptions of
the writers, or the traditions of the age.

Objections
from want
of distinc-
tion be-
tween the
different
parts of the
Bible.
Difficulties of this kind in many instances, especially as regards the Old Testament, are raised into
importance to Christians only from the common want
of due discrimination as to the distinct object and
character of the different portions of the Bible.

If, however, we look with a more discerning eye
to the nature of the contents of the Old Testament,
in the first instance we find a record of older and
imperfect dispensations, adapted, as they were addressed, to the ideas and capacities of a peculiar
people and a grossly ignorant age—a law of " carnal
ordinances " and sabbaths, specially founded on that
peculiar cosmogony which we now know to be untenable; physical influences, temporal and national
retributions.

But the more perfect and universal religion of
Christianity, if in its first outward manifestations
accommodated to the convictions of the people among
whom it originated, yet in its essential characteristics
and more full development to the rest of the world
as set forth in the Pauline Epistles, while it expressly disclaims the peculiarities of older dispensations, exhibits characteristics of a higher, more
comprehensive, and spiritual character; professedly
appealing for their acceptance to the principle of
faith, not of *sense*.

Christianity independent and spiritual.

If we look to its more special doctrines; as *regards
the Divine nature*, we may observe that a physico-theology supplies *no such idea of the Deity* as can offer

Doctrines of the Divine nature.

any antecedent contradiction to the representations of His nature and attributes (necessarily more or less anthropomorphic), or to the spiritual mysteries of redemption and a future state, in the form in which they are announced in the New Testament.

Of the nature of man.

And as to *the condition of man,* the language of the Christian doctrine represents it not in reference to any material, metaphysical, or moral hypotheses, but to peculiar spiritual principles. It makes the spiritual man a distinct being from the natural, — *" a new creation; "* * and is engaged not in tracing *physiologically* the *origin* of the human race, but in pointing to its *future* destiny : not in detailing the *sources* of man's *infirmities,* but *in providing the* RE-MEDY. It does not dwell on external events in any physical detail, but always with a doctrinal application, or in a spiritualised meaning. Its *essential* design belongs to *spiritual* things; its relation to external and *physical* things can be but subordinate ; and of the proper objects of its revelation we may truly say with Bacon, *" Dignius credere quam scire."* †

Christi-

Thus we need be in no trouble " how to place our

---

* Gal. vi. 16. ; 2 Cor. v. 17.   † De Augm. lib. ix. c. 1.

theology " among physical systems ; nor need it fear
any speculations as to the inhabitants of other worlds or possible revelations granted to them.   It has no concern with chronology, astronomy, or cosmogony (least of all, that of the Judaical law); with the nebular origin of the planets, or with the development of successive races of organised beings in them or on the earth; with the myriads of ages which mark the antiquity of the world, or the date of man's origin upon it; or with the question of his derivation from one stock or many*, or the origin of civilisation.   It leaves these questions to be guessed at as they may, or investigated on philosophical principles so far as they can be.   Its peculiar aim is entirely different and independent: its objects *belong to another order of things ;* and its representations of them are avow-

---

* While writing this Essay, I have received a copy of a pamphlet, circulated privately and anonymously, in which an able and learned writer, evidently a strict upholder of inspiration, endeavours to show by elaborate, critical, and philological investigations, that the Bible distinctly sanctions and asserts the idea of the primeval existence of other races of men besides the family of Adam.   It also includes a defence of this belief against its supposed unfavourable influence on the doctrine of original sin, on the ground that Adam might be spiritually the "representative" of the human race, without being *physically* their progenitor, just as Christ was without any such physical relation. ("The Genesis of the Earth and of Man," &c. printed for private distribution, July, 1854.)

An expansion of this pamphlet into a volume has now been published, *edited* (but *not written*) by J. S. Poole, Esq. of the British Museum.   It includes a novel attempt to reconcile Geology and Genesis, of a kind even more *visionary* than any of its precursors!

edly *not the realities*, but only their *images ;* they can
be seen by us only " δι' ἐσόπτρου ἐν αἰνίγματι *, — by
means of a mirror and in an enigma," in our present
state ; while it holds out a future, when " we shall
see face to face, and know even as we are known."

---

* 1 Cor. xiii. 12.

ESSAY III.

ON THE
PHILOSOPHY OF CREATION.

THE PHILOSOPHY OF CREATION.

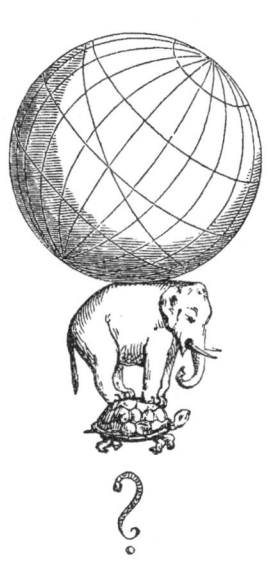

INTRODUCTION.

THE question of "creation," — whether in the higher sense, of the first origination of the material universe, and of all physical causes, — or in the secondary and more accessible meaning, of the earliest history of the cosmical arrangements of stellar or planetary systems, — and more particularly of our own globe, of its physical revolutions, and of the successive introduction of new forms of organised

Prelimi-
nary re-
marks.
Interest of
the ques-
tion.

life on its surface, — has at the present day excited an unprecedented degree of general interest.

Various grounds of discussion.

The discussion which has been called forth has exhibited the greatest variety of tone and character. Though by some the subject has been viewed in a sober philosophical aspect, yet by others it has been made the subject of hypothetical speculation; sometimes running off into what must be deemed very fanciful imaginations, or occasionally involving metaphysical ideas, carried out into various stages of abstruse and even mystical speculation. By others, again, it has been taken up on religious grounds, or mixed up in various degrees with the influence of a theological creed, to which speculations of a more scientific kind have been deemed hostile.

Religious bearing of the subject.

That the subject has a direct connexion with such higher considerations is manifest; but this by no means implies such a confusion of ideas as would vitiate the claims of independent philosophical inquiry, or sanction the attempt to found conclusions relative to the physical order of things and the structure of the material world on a basis totally alien from that of *inductive science*. The truth is, the same observations will in a great degree apply to this question which were offered in a former Essay*; and the strange speculations which have been sometimes broached respecting it have probably originated in a great degree in the want of clear appreciation of the distinct grounds on which our convictions of scientific and of religious truth respectively should be based. And until these distinctions are properly drawn in the mind of the inquirer, it will be to little purpose to discuss the details of controversy.

It will be allowed, then, in conformity to principles before laid down, that on purely physical and inductive grounds it is fully open to us to inquire how far science can legitimately conduct us towards some indications of the *mode* in which, and the *secondary processes* by means of which, the first establishment of the existing natural world may have been worked out, and, limited strictly to the tenor of recognised natural analogies, to speculate on the probable order of evolution of the earliest rudiments of life.

In proportion as such speculations have a tendency to impress a conviction of truths of a higher order,

Properly a subject for physical inquiry.

* See Essay II. § II.

it is the more necessary to the force and validity of such inferences that the *evidences* on which they rest should preserve a free, independent, and un-prejudiced character, and should not involve an assumption of the points to be proved.

Meaning of the term " creation."

It must also be borne in mind, that we have by anticipation, in several parts of the foregoing Essays, discussed more or less in detail several topics which have a material bearing on the present subject, as referring to the past history of the world, and those stages which it has passed through in the process of formation ; all which are essential to any *physical* view of the nature of its *creation ;* that is, its history, *so far as we can trace it,* towards its first origin.

The very use of the term " creation " may in-deed be supposed to point to associations of a higher kind, which are altogether beyond the simple scientific question. But in a philosophical sense, it should be carefully borne in mind, that if that term be employed, we can regard it as no more than an expression of our ignorance as to the *mode* of the first origination of the material world; while, as to the secondary points connected with that ques-tion, we may always look to the further enlightenment which discovery will continue to throw upon them.

Even in earlier times, and under very erroneous systems of philosophy, we trace some recognition of these sounder principles. Amid the many extravagancies and radically erroneous principles of the philosophy of Descartes, we may yet recognise its principal merit in that, in an age when *metaphysical* abstractions and causes alien from natural induction were universally resorted to for the solution of *physical* phenomena, his theory of vortices (imaginary and fallacious as it was) yet at least referred to *conceptions and modes of action of a properly physical kind :* and it is remarkable, that in conformity to the same broad principle, he likewise distinctly upheld the notion of the origin of the existing *organic* world, evolved according to a series of regularly adjusted laws out of its primitive elements ; though like some other *very prudent* philosophers, from a salutary apprehension of popular odium, he was unwilling openly to avow such an obnoxious tenet, and in his popular writings professes to think it more likely that the whole was created at once as it now stands.

In such discussions the first requisite ought to be

Physical evolution.

Want of

sound phi-
losophical
principles.
that the inquirer should make up his mind on what ground he is intending to proceed, whether physical or metaphysical, inductive or mystical; yet simple as this consideration is, it is too commonly but little thought of, or even purposely disguised or kept out of sight. If we profess to go on the sole ground of philosophical analogy or rational conjecture, our way is clear ; every consideration not connected with such a view is inadmissible in science and must be

Sole pursuit
of truth.
peremptorily discarded. If once any considerations of a kind foreign to the simple inductive view of truth are suffered to intrude on the conclusions of impartial reason and dispassionate conviction, there is an end of all philosophy. Nothing can be more fatal to the pursuit of truth than a disposition to look at conclusions not according to the *evidence* adduced, but the *purposes* they may serve or the authority by which they may be countenanced.

Let the advocate of other objects (excellent and valuable in their way) consistently profess and follow them up, and he may be eminently useful and estimable : but let him not mistake aims and confound purposes of different orders. Let him not make pro- fessions of philosophy, and then abandon the character of a real interrogator of nature, by yielding to other influences and aiming at other objects. Let him not assume the badge of science and yet serve utility, or bow to authority,—or set up strange gods in the temple of truth.

Yet we find, in fact, many instances of writers professedly treating subjects of a philosophical kind, whose arguments are too often those of partisans rather than of philosophers; who betray too much of a determination at all hazards to support a " safe " hypothesis and repudiate and discredit an obnoxious one, while they are ready to adopt any evasion, any form of ambiguous mystification, to screen themselves from the reproach of being supposed to hold opinions opposed to the popular voice. In no instance have these remarks been more extensively exemplified, than in the discussion of questions relating to the view of " Creation."

In carrying on the present inquiry with special reference to some of the theories started at the present day, I propose, first, to consider briefly the general amount of information which can be regarded as bearing on such a question, furnished by those branches of science most directly connected with it, and then to offer some remarks on the general character of the reasoning raised upon that evidence, the kind of conclusions we may safely deduce, the kind of hypothetical speculation in which, if so disposed, we may legitimately indulge, and the extent to which any real conceptions can be carried of a subject which, in its entire compass and highest meaning, must necessarily be beyond the reach of positive investigation, or even of human comprehension.

§ I. — THE EVIDENCE DERIVED FROM GEOLOGY.

IN any question as to the origin of the world pursued on scientific grounds, the sources to which we can look for any positive or substantial information, must, in an especial degree, be those opened to us by geology and palæontology; and of some of the most material facts and admitted theoretical opinions in these departments, bearing on the question, it will be necessary to take a cursory review.

Sources of information. Fossil remains.

When we trace backward, by the light of fossil remains, the succession of varied forms of organised existence which have tenanted the surface of our globe during the incalculably vast periods of past time, the fact of their presenting apparently very different characters in different epochs naturally led geologists and naturalists to speculate on the question, whether those variations could be reduced to anything like *a determinate order or law of succession;* and probably the most prevalent opinion has

Theories of progression and non-progression.

z

been that, at least in a general sense, there has been a succession in the order of *progress* or *advance* from lower and more simple, towards higher and more complex forms of structure and function.

More recently, however, this has been much disputed. Not only have particular instances, supposed to invalidate this law, been brought forward as demanding certain modifications in the statement of it, but the entire principle has been contested and positively denied. And those who have pursued the inquiry (*restricted merely to the question of fact*) have been ranked under the two schools of Progression and Non-progression.

It may, indeed, be fairly questioned whether such distinct designations, applied without qualification, can be fairly supposed to characterise any parties in the scientific world, or whether we may not rather regard the differences as of a more limited character. But at any rate, a very brief summary of the principal arguments on either side will materially conduce to our object.

Arguments for non-progression. On the one side, Sir C. Lyell has supported his views by insisting on the merely *negative* character of the evidence we possess as to the non-existence of many species in early ages *; on the *absence* of any indications in some of the earlier formations, from which we can form an idea of what the entire flora and fauna of those periods really were. More precisely, those formations being wholly marine, we have only evidence of the marine organisms, and can expect none of the contemporaneous land productions; and marine plants and animals are confessedly always of lower organisation. Throughout the long period of the Silurian formations, he contends that we have little evidence of any advance or progress in the scale of organisation. The fishes of the coal formation are of higher organisation than any existing species — in fact, combining reptilian characters with icthyic; and even up to the later formations he conceives that there is but little indication of any real advance in character.

In general, that the remains actually preserved in any formation furnish us with no certain standard or adequate representation of the entire existing state of the organic world at the time, is evident from the very partial, local, and accidental manner

---

* Address to Geological Society, 1851.

in which those remains have become imbedded ; their preservation has been the *exception* rather than the *rule*. We ought not to expect to find evidence of the possible multitudes of co-existing species which might be so circumstanced that their remains were never likely to be thus embalmed for our instruction.

No proof of inferiority of earlier fauna.

On such grounds, then, it is concluded that we have no real proof of the *general inferiority* of the organic world in the earlier periods, or of any superiority, at least through a long course of succeeding ages ; nothing to lead us to trace backwards any determinate series which points to a primary rudimentary condition, or to the origination of a *more perfect* state of things out of a *less perfect*.

Arguments for progression.

On the other side, the argument which has been so ably sustained by Professor Owen*, turns chiefly on the *positive* evidence supplied by the comparison of those organic remains which *are* preserved to us in any one formation, with those in another, when, in fact, the same probabilities of preservation must be supposed to have subsisted, and yet the remains

---

* See Quarterly Review, Sept. 1851.

of each exhibit such marked relative characteristic differences, that thus, he conceives, the inference of considerable progressive advances may be fully justified. But his main conclusions are supported, not so much on general arguments, as on minute anatomical comparisons; and such comparisons, in at least very numerous instances, present undoubted marks of physiological changes, clearly progressive according to the order of higher organic development. *Comparison of preserved forms in different periods.*

But in some instances, it is further urged, we have a more positive ground of inference of the real absence of particular species, when we find other forms closely allied which are, by analogy with the rest of the system, fairly considered as the analogues or *representatives* of the missing species. We infer, as it were, the absence of the *principals* from the presence of their *substitutes*. The argument, of course, depends on the force and correctness of this analogy, and the general evidence of such a system. *Analogies of species.*

With regard to the absence of terrestrial remains, it is considered by some as evidence that, in the earliest periods, the sea covered the whole surface of the earth, or nearly so. But there must have been *Marine formations.*

some elevations to be washed down, and form the deposited beds.

Again, if progression be supposed, it is an admitted part of the law, that in the *lower* forms, whether of animal or vegetable life, the change has been, throughout all the series of eras, always much less than in the higher, and that some of the lowest forms are persistent, or nearly so, through all formations. Thus it would be in proportion as we ascend to the higher classes, that any marked signs of change or improvement might be expected.*

There is one material consideration the force of which, *on any view of the question of progression*, it is impossible to overlook — the fact of the *central heat* of the earth, with its undeniable consequences.† A hot body in free space must cool; and if now cool at the surface and hot within, the earth must have cooled from a hotter state, and must once have been

---

* Ehrenberg in his " Mikrogeologie " has lately given the results of a most elaborate and extensive set of observations on the microscopic fossils of all formations, and gives as the general result, that these minute infusorial species, unlike those of higher orders, evince little proof of change in type, in relation to the age of the deposit. He finds the same genera, and sometimes the same species, extending from the most recent formations to the carboniferous, and in some instances even to the lower Silurian. Geol. Quarterly Journal, No. 42. p. 89.

† See especially Mr. Hopkins's Address to the Geological Society, Anniversary, 1852.

intensely hot; and, by the same rule, once in fusion, or in vapour.   Here, then, there *must* have been a series of *progressive changes;* and if the cooling process *had not reached* its present apparent state of equilibrium *before* organised life began, it would be a natural consequence that some marked changes in animated nature could not but have accompanied those changes in temperature, and have followed a like determinate order.   Whether the present equilibrium had been attained before the commencement of life may still be a question; yet, considering the enormous length of time through which organisation has certainly existed, it is difficult not to suppose that some part of the series at least must have reached back into the period of perceptible cooling, so that influences of terrestrial temperature may have been not without their effect on the changes of species.

On the other hand, these considerations may admit of qualification, if we should agree with those who ascribe the terrestrial heat wholly or partially to other causes than simply the remains of a primitive high temperature.   If there be admitted, for example, any internal cause of combustion capable of being more or less excited from time to time, any

Supposed local causes of heat.

effects of a progressive kind must be greatly influenced; and the advocates of the chemical theory
of volcanic action allege much probable evidence in
support of such an idea.*

Again, possible changes of temperature from external cosmical causes might interrupt those due to
progressive cooling. Nor need we dwell upon those
more immediate local causes of change of climate in
the variations of physical feature in the continents
and oceans, universally recognised by geologists,
which might interfere greatly with any general progressive change of temperature in particular localities.

Influence of
tempera-
ture on or-
ganic life.

Another question might arise:—Are the organised productions, at epochs when a tropical climate
prevailed in districts now temperate or cold, upon
the whole of lower organisation? or are we to infer
that a hotter temperature is less favourable to the
evolution of higher forms, or a colder more so?

In one word, we are not certain as to what were
the *successive order* of changes in the temperature of
the earth; nor if we were, could we thence argue

---

* See Daubeny on Volcanoes, p. 430. (Ed. 1826.)

what kind of successive changes in organised life should be expected to accompany them.

Upon the whole, when we carefully examine all that has been alleged on either side, we cannot deny the evidence, in some sense, of progressive changes on the one hand, though even throughout some long periods we may allow the amount of change is small, and the apparent amount greater than the real. All advance is at the utmost extremely slow; and with the progress of discovery, there is continually increasing reason for believing it slower than has been imagined, and that a high type of organisation prevailed in epochs much more remote than has been supposed; while almost every fresh discovery tends to push backwards the boundary which seemed to mark an inferior order of things into remoter depths of primeval time. In fact, the discussion of the question of progression or non-progression is perhaps less valuable on its own account than as it leads to a more searching review and analysis of the ground on which all geological reasoning proceeds.*

General bearing of the question of progression.

---

* On this point see some admirable remarks in Lyell's " Manual," 5th Ed., 1855, p. 457.

Force of
negative
evidence.

We are thus led to consider generally the proper force of *negative evidence* as such: in other words, that any inference made from the *mere absence* of discovered instances must be *essentially dependent on concomitant circumstances.* Negative evidence by itself is simply neutral; but it acquires a different character and force according as other arguments concur with it or otherwise. Negative evidence is *strong* in proportion as we may be able to show from circumstances a high probability that instances would be found if they existed, or a high probability, from other analogies, that they did not exist. *Non-appearance* would here be nearly tantamount to *non-existence.* But negative evidence is *weak* in proportion as we may be able to show from circumstances, a probability that instances would *not be observed* even though really existing; and *still weaker*, if analogy should render it likely that they did exist. *Non-appearance* would here be no presumption, even, of *non-existence.*

Origin of
life on our
globe.

As to the great question of the *first origin of life* on our globe, geology can give us very little, if any, information. An *azoic rock* is no necessary proof of an *azoic period.* Animals may have lived and flourished abundantly where, from peculiar causes, none of their remains might have been imbedded. Analogical considerations may avail to guide conjecture to a certain extent. But the earliest forms AS YET *known* are *not* of the lowest organisation. And if we descend to the so-called primary rocks, or to those called metamorphic, it is clear, whatever remains they might have included must necessarily have been fused and burnt up.

But in relation to the question of the absence of organic remains, we must not omit the consideration, that though organic *forms* might be destroyed, yet the presence of that constant element of animal life, phosphoric acid, — incapable of dissipation by heat, — would be a proof that animal remains had once been imbedded, if detected by chemical analysis. Such analysis, however, being attended with great difficulties, we owe, perhaps, the first intimation of the fact to the ingenious suggestion of Dr. Daubeny, by growing plants in the pulverised· soil and comparing the proportion of phosphoric acid in the produce with that in the seeds; and in this way a minute quantity was detected in the Bangor

Azoic formations.

and Llanberis slates: proving the existence of animal life in those apparently azoic formations.

In the inductive prosecution of the question of the progress of life, the first object would be to endeavour to determine the *law* to which the order of *succession* of species in different epochs may be found to conform.   The first and most simple idea of a direct advance in successively higher organisation in one line, from the lowest zoophyte up to man, as we advance through geological ages up to the present time, is now acknowledged to be untenable ; but what is the real order which we are to substitute for it, is not so easily apparent.

The assertion is often dwelt upon with a very mistaken emphasis and importance, that a particular species is *highly organised,* when the real point of distinction should be, not its *absolute* but *relative* degree of development ; when the question is, not whether its structure is actually of a complicated kind or exquisitely adapted to the conditions of its existence (of which in no case is there any doubt), but whether it is of a *higher* or *lower* grade *relatively* to other creatures of a corresponding class in other formations.

And, again, it is a point of material importance, Scales of advance. but one in which few writers are agreed, what we mean by *advance* or *progress*, or what really constitutes a *higher* or more perfect organisation.

It would rather seem that each species is higher in some respects, and lower in others ; or that there are *many scales of perfection in different respects*, runing, as it were, *parallel* with each other; and that in defining the degree of elevation of any species, we must take into account the position it occupies in the several different scales jointly.

Among existing animals, it is now generally allowed that the arrangement of species in the scale of organisation is nót that of simple ascent in any line, or even in several branching in any one direction; it is more properly compared by Professor Owen* to a " net-work ; " — every species being connected with others by a variety of ramifications, and not simply by ascent or descent in a scale.†

---

* Lecture, British Association, Liverpool, 1854.

† The same difficulty in the definition of *higher* or *lower* organisation has been felt also by other naturalists.   That species may be higher in certain respects and lower in others, is also dwelt upon and illustrated by Professor Pictet, " Traité Elémentaire de Paléontologie," &c.   (See Geol. Quarterly Journal, No. V. p. 50.)

And Oken, in proposing his new scheme of classification founded on the modifications of the organs of sense (Physio-Philosophy, § 3065.), con-

And as to the law of changes of species in *past epochs,* in the very imperfect knowledge we at present possess of it, it is at least clear that, so far from a regular *advance* from lower to higher forms, in many instances there appears rather a deterioration and degradation of character in the progress of time towards the existing state of things.   But what seems most material towards the probable ultimate enunciation of a more true and general expression of the case, is the law of *combination and separation* of characters; that is, a combination of the characteristics of several species, or even genera or orders, in the same individual in one period, to be developed separately in different species in a succeeding era; and this in such distribution as to present appearances of advance in some respects, along with degradation in others; as if, in the functions of vitality, the principle of " division of labour " had been gradually introduced.   Of this combination of characters in an individual species, examples are familiar

Marginal notes:
Not a simple advance from lower to higher.

Combination of characters.

Afterwards separated.

Examples.

---

fesses the difficulty of arranging animals on any satisfactory principle in the relation of *higher* and *lower* organisation. (§ 3561.)

In his own system, he expressly notices that, while each class stands above another, yet *in each* the lower animals are inferior to the higher animals in the next below. (§ 3582.)

to every student of geology. The sauroid fishes generally furnish obvious and abundant instances; and we may take as a single case which has been much dwelt upon, the *jaw* of the Asterolepis; in which a row of small fish-like teeth are combined with another set of the large reptilian form, or, as Mr. Miller so forcibly expresses it, we find " the crocodile lying intrenched in the fish; " at the same time its general organisation was of the inferior type of the cartilaginous fishes, having external plates resembling those of the recent Lepidosteus and Polyptemus, and a spiral coprolite indicating a visceral structure like that of the ancient Icthyosauri and the existing rays and sharks."[*]

Indeed, it might perhaps even be conjectured to be more like the general law, that this kind of *combination of the characteristics of higher, with those of lower, classes*, might be the distinguishing feature of all the earlier stages of animal life; and that the higher we ascend in time, the more we might expect to find types combining characteristics of several

---

[*] Footprints, &c. 80. 104.

(perhaps of all) classes, thence afterwards to diverge in distinct directions.

Principle of the continuity of geological phenomena.

One of the most material points in the whole inquiry relates to the question of *continuity of character* observable in palæontological indications throughout successive formations. To a great extent such continuity is on all hands admitted as marking at least large portions of the series of changes presented to us ; but an important question arises respecting *interruptions apparently occurring* in that order and gradual succession of forms, on which considerable difference of opinion has prevailed. On this point, then, we must make a few observations.

Successive subdivision of all formations.

In the first place, the general tendency of all geological discovery has been, and continues to be, to break up large divisions into smaller, to obliterate sharp lines of demarcation by subordinate gradations ; to subdivide formations ; to trace intermediate deposits, lost perhaps in one locality, but detected in another ; and thus its course continually tends to fill up breaks, to render the series more and more connected, and to confirm the belief in a real continuity of geological phenomena ; though we may *as* yet, be very far from realising it in all instances, or throughout all the series of changes.

All geology is full of instances of such progress. Instances. It is not many years since the whole mass of rocks below the old red sandstone, and above what was called the primary, was confounded together under the common name of " grauwacke." All the strata, again, above the chalk were alluvium or diluvium, London clay and fresh-water beds.

But we have now, in the one case, through the combined labours of Sir R. Murchison and Professor Sedgwick, the vast mass broken up into the well-marked series of the Silurian (including the Cambrian) rocks, with their several subordinate formations and accompanying beds; while some of these are again in process of undergoing still further analysis; as in the researches of M. Barrande : again, in the other case, the labours of Sir C. Lyell and his later coadjutors, in the first instance reduced the chaos into order, by the grand divisions of Eocene, Meiocene, and Pleiocene; in their turn subsequently broken up into an increasing number of minor distinctions of older and newer Pleiocene, Pleistocene, and Postpleistocene; no doubt, eventually to be still further marked out by yet more minute shadings of difference in epoch; and thus indicating in every instance a more gradual succession and closer approximations in the affinities of species.

Throughout all formations, the grand truth to which every accession of geological discovery bears witness in a more remarkable manner, is the principle of unity of plan continually exemplified in all the varieties of organic structures disclosed. Even the most seemingly monstrous and incongruous forms of animated existence in past times are all, without exception, constituted according to regular modifications of a common plan, and with parts, organs, and functions related by the closest analogies to each other; so that no sooner is a new specimen detected than it immediately finds its proper position in the scheme of nature; no sooner is a new form discovered than it is instantly assimilated with some known type, and found to hold an assignable place in the system. Whether a given organic fossil (as in some instances in more recent beds) exhibit characters differing from some known form only as a variety or sub-species, or whether (as in earlier cases) it present features unknown to any existing genus or order, or (as in other instances) offer conditions in any degree intermediate, still in all cases alike the remark-

able point is always, that a place and a name can be
immediately assigned to every new form as it presents itself; and this too invariably in such a manner
that it either tends to supply a link in affinity be-   Connexion
tween orders of beings already related, or indicates   in affinity
of forms.
some new and unexpected point of analogy.   There
is never any deviation from system and regular
plan ; we never light upon a fossil centaur or palæozoic mermaid; there never occurs any junction
of heterogeneous members, any real departure from
type and system.   The invariableness of the results
through such enormous series of ages cannot but
impress the mind, when duly considered, with the
highest idea of the preservation of continuity.

Throughout all the most recent formations, indeed,
we find a continuous series of allied species, and a succession of organised structures, in a chain absolutely
unbroken, and marked only by the minutest specific
differences in its successive links, down to forms now
existing; and as this is carried backwards through
countless ages, by degrees we find fewer features of
the present, and more of the past, and even come
to whole genera, and orders of extinct races coexist-
ing with some which have survived them.   But in

some instances, especially in the more ancient formations, the series of forms present a more fragmentary
appearance.   At intervals in the course of this series
of close and continual connexion, there are real or apparent interruptions of greater or less magnitude, in
which the immediate affinity seems broken off between the species characterising one formation, and
those nearest allied to them in the next formation.

Some appa-
rent breaks
in the
series.

The case of apparent breaks or discontinuities between one great group of formations and another is
often alleged as one main difficulty in the way of
any theory of continuity; and this is evinced in several marked instances in an apparent interruption,
not merely in species, but even in genera; that is,
though through considerable ranges of closely con-
secutive deposits the transition of species takes
place by insensible gradations, yet, at length, we
come to a broadly marked separation of that group
from another, where not only the species, but even
the genera, disappear and are replaced by others.

Examples.

To take a single example; one of the most remarkable of such apparent interruptions is, that
marking the boundary between the Permian beds —
the highest of the older group of fossiliferous rocks —

and the Trias, or lowest of the next series; and
to which respectively the terms Palæozoic and Mesozoic have been applied, as evincing an apparent
great change in the organic life which prevailed
in those respective periods.

Sometimes a stratum containing, perhaps, abundance of fossil remains of a particular class or epoch,
is succeeded by a great thickness of deposit totally
devoid of organic remains ; and then, in the next bed
below this, organic remains shall again occur abundantly, but of totally different *species*, or perhaps even
genera, from the last, which, it is contended, indicates
the occurrence of a long interval of time after the
destruction of the former, and followed by the introduction of the latter kinds of beings, without any intervening links in the chain of existence appearing.

Thus Mr. H. Miller dwells with peculiar emphasis
on an instance of this kind occurring in Orkney, where
there is a bed of the lower old red sandstone containing an abundance of fossil fish; greater (according
to this author) than in all other formations together;
in which the celebrated Asterolepis was discovered.
Superimposed on this are other beds of sandstone
reaching to 1500 or 1600 feet of thickness, in which

not a single organic fossil has been observed; above which are beds containing totally different species. Again,* he argues on the fact that in the Silurian system, fossil fish (at the time he writes) have been discovered only in certain beds in the upper division; while the lower, more than 3000 feet in thickness, are destitute of them, and below these we arrive at remains of a different character.

Objections against continuity.

These phenomena, and others of the same kind, have been the subject of considerable dispute; and the opponents of the doctrine of continuity, chiefly on grounds which it is difficult to recognise as connected with those of true science, and often in a tone still less reconcilable with its proper spirit, have been fond of triumphing in these facts, as if they inflicted a fatal blow on the views of their opponents. It is not my intention here to descend into any such polemical disputes. I merely proceed to the philosophical consideration of how far any such phenomena (granting the representations made of them as accurate) really affect the question of continuity. It

General view of the evidence.

will here be important to recur to the consideration

---

* Footprints, pp. 106. 114.

of the nature of negative evidence ; and to observe that the *absence* of all organic remains in a particular formation or deposit is *no proof whatever* that animal life did not abundantly exist during the whole period of that deposit, but merely shows that local and other causes did not favour the imbedding and preservation of their exuviæ.

The popular apprehensions as to the nature of geological events are often very inadequate and confused ; and it is a point apt to be overlooked, that the terrestrial remains in all formations are merely *indications of what was the state of the* LAND *left us by the* WATERS, whether of the ocean, of rivers, or of lakes. Such remains were only occasionally imbedded — "rari nantes in gurgite vasto" — and thus afford no adequate representation of terrestrial life. Even marine remains are far from affording a complete memorial of the inhabitants of the ocean. At all events, it is a hazardous process to frame *theories* on the *absence* of such remains.

Aqueous deposits.

Exceptions may, indeed, be conceived in cases where, instead of being formed by sedimentary deposition, a tract of land, with its plants and animals, may have been *submerged by subsidence ;* here a fairer

Submergence.

representation of the whole fauna and flora might be expected: but, perhaps, very few, if any, instances affording good evidence of such a process have been clearly established.

All deposits local and occasional only.

Again,—all the formations which geology has traced were simply *local* and *occasional* deposits, extending sometimes over a greater, sometimes a smaller area; and going on at one time, and ceasing at another. Equally local, too, was the diffusion of organic forms.

Professor E. Forbes* has justly observed in his able comment on the labours of M. Barrande in the Silurian formations, " Thus early in the world's history do we find the partitioning of the earth's surface into natural history provinces; more and more evident does it become every day that the old notion of a universal primeval fauna is untenable." And to the same effect I must refer the reader to some profound observations of Mr. Darwin †, into which my limits alone prevent entering here at the length they deserve.

At any one epoch deposits might be going on with

---

* Address, Geological Society, 1854, p. 33.
† Geology of South America, p. 105.

more or less regularity during a certain period, en-
closing remains; afterwards, during another period
equally long, or much longer, a *total cessation of deposition* might take place, owing to changes in local
condition; or, again, at one place *deposits* might be
going on along a shore, while over a vast region,
away from the waters, species of terrestrial animals
might be flourishing in profuse variety, not one
fragment of whose *remains* might ever be *washed
down* or *imbedded* either in lacustrine or marine beds;
in a way, in fact, exactly analogous to what is going
on at the present day.

Again, it is alleged that the change from one
great group of formations to another, at least in
several marked instances (as *e. g.*, in passing from
the palæozoic to the mesozoic period), was marked
by the occurrence (according to some) of a "great
convulsion," or at least of very extensive changes
in the physical order of things, of which the
condition of the strata sometimes bears striking evidence. Now, granting such changes as great as
the catastrophist may imagine, it is surely a most
unreasonable inference that these changes were
such as to *destroy* all the species existing during the

Alleged
break be-
tween the
earlier
formations.

previous formation, and thus to leave the surface wholly untenanted until a new order of things supervened, and a totally new introduction of life took place.

Lapse of time between successive deposits.

But granting that between the periods of formation of the upper and lower groups referred to, great changes in physical arrangements took place, it would be far more accordant with all reasonable analogy, in proportion to the magnitude of those changes, to allow a corresponding lapse of *time;* which being unmarked by any depositions, giving evidence of its duration by the successive changes they might exhibit, would necessarily remain to us a blank; a period which the advocate of natural causes may, with just as much probability and confidence, assert to have been enormous and incalculable, as the catastrophist can maintain it to have been brief and spasmodic.

If an interval of unknown and incalculable length intervened between two recognisable formations, and during all this vast time circumstances did not allow the imbedding of any characteristic exuviæ, it would be utterly vain and futile to assert that there was necessarily any breach whatever of *the law of con-*

tinuity; or to affirm that, during the whole of this enormous period, of which we are, from the conditions, precluded from knowing anything, all the species of the earlier epoch were not continuously existing and as slowly changing (by whatever means or law) to others more and more different, along with corresponding slow changes in physical conditions, until at the period when things were such that remains were again deposited, the whole character of the fauna had changed in the manner observed.

In a word, in all those geological periods during which we *can* trace a continuous and gradual succession of formations without marked or violent interruptions, there we invariably find a like slow and gradual change of animated life, proceeding by small modifications of *species*, until, at length, comparing the extremes of the series, whole genera may be changed. If, then, in certain other cases, we find apparent *interruptions in the order of species*, apparent breaks in this orderly succession, or between such deposits of so different a character, periods intervening, during which we see that great changes or disturbances were in progress, as we must infer that those changes went on by the regular operation of

Analogy
with those
cases when
deposits are
continuous.

physical laws, exactly as in the cases in which we
have uninterrupted evidence — so, by parity of reason, we must infer that the like gradual and regular
changes of species went on during those periods,
though all its intermediate links and steps are lost to
us, and only the extreme terms are preserved.

Non-fossi-
liferous in-
tervals in
one and the
same form-
ation.

We have one striking proof of this, in the
fact, perfectly familiar to geologists, that in many
formations we frequently encounter a thin layer
of their characteristic fossils, upon which succeeds a
large, and sometimes an enormous, thickness of the
same deposit, *wholly destitute of organic remains;*
after which again occurs another thin layer *full of
them;* and this sometimes *repeated* more than once
in the same formation; a distinct proof, therefore,
that while these beds, destitute of *all indications*
of animal life, were being deposited (which must
often have been a period of great length), animal
life was still really going on in full intensity and
variety, though from local causes no specimens of
it were imbedded; yet its continued existence was
evinced again when the upper fossiliferous bed came
to be deposited.

Recapitula-
tion.

To recapitulate : — The argument from the known

to the unknown is clearly this : in one instance we find two different epochs, at which species or even genera exhibit a wide difference : of the interval between these we have, however, continuous evidence showing that, during this vast period, species have gone on changing by insensible gradations, until, taking its two extreme points, they exhibit that wide difference alluded to. Again, in another instance we find two different epochs, at which species or even genera exhibit a like wide difference. Of the interval we know nothing : we have either no intermediate beds, or an azoic mass. The obvious inference from analogy is, that that interval was probably as long, and was marked by as gradual changes, as the former, though circumstances have prevented their being exhibited to us.

If, then, we find a bed containing certain species, and then superimposed on it another containing forms not only specifically, but even generically or still more widely different, instead of a real hiatus, an interruption, a destruction, and a sudden reproduction of life, the fair inference would be the occurrence of an indefinitely long interval of ages, during which, indeed, no fossiliferous deposits took

Break in succession of species only shows long interval of time.

place at that locality, but during which the slow progressive change of species went on, until whole genera
were different, and then a deposit took place in which
some of these latest remains were imbedded. *The
wide organic difference between two contiguous beds
would only mark the longer interval of time between
their deposition.*

In confirmation of the ideas thus suggested, I
have great satisfaction in citing the testimony of
two very distinguished men, each delivered from the
chair of the Geological Society. The first I will
quote is a single passage from the anniversary
address of Mr. Horner, who, amid a variety of other
able remarks bearing on the present subject, observes, " By whatever names we designate geological periods, there appear to exist no clearly defined
boundaries between them in reference to the whole
earth. Such a marked line may be seen in particular localities, but every year's experience, and our
more intimate acquaintance with the phenomena
exhibited in different countries, and with the distribution, structure, and habits of animals and vege-
tables, teach us that there is a blending, a gradual
and insensible passage from the lowest to the highest

Views of
Mr. Horner.

sedimentary strata, particularly in respect of fossil remains. The terms we employ to designate formations can only be considered as expressing the general predominance of certain characters, to be used provisionally, as a convenient mode of classifying the facts we collect, whilst that knowledge is accumulating, which in after ages will unravel the complicated changes that belong to the successive periods into which the history of the structure of the whole earth may be divided." *

The second opinion which I have to quote, is that of the late Professor E. Forbes, who says, " I am one of those who hold, *à priori*, that all gaps are local, and that there is a probability, at some future time, of our discovering gradually, somewhere on the earth's crust, evidence of the missing links. All our experience and knowledge, theoretical and practical, warrant the affirmation that, at every known stage of geological time, there were sea and land. Even those who believe in a primeval azoic period will hardly sanction the supposition that there has been any repetition of azoic epochs

Views of Professor E. Forbes.

---

* Address to the Geological Society, 1847, by Leonard Horner, Esq., President, p. 22.

since the first life-bearing era commenced. And if so, and if there were always sea and land since the commencement of the first fossiliferous formation, we are warranted in assuming that both earth and water had their floras and their faunas."

" All geological experience goes to show that, whenever you have a perfect sequence of formations accumulating in the same medium, air or water, as the case may be, there is, if not a continuance of the same specific types, a graduated succession and interlacement of types, and of the facies of life-assemblages ; even as, on the present surface of the earth, the faunas and floras of proximate provinces intermingle more or less specifically ; or, if physical barriers prevent the diffusion of species, assume, more or less, one general facies. This passage by aspect and type of one stage in time into another, is but scantily indicated at present in the uppermost manifestations of palæozoic life, and the lowermost of the mesozoic. The missing links will sooner or later reward the diligence of the geological explorer."*

* Proceedings of Geological Society, Address, 1854, No. 38. p. 78.

The author, however, conceives that, notwith-standing this general unity, there are features in the distribution of organic existence in time, which seem to indicate some real law not as yet recog-nised; especially in the instance referred to, of the slighter connexion in sequence between the meso-zoic and palæozoic periods; and he proceeds to suggest an explanation by applying a new theo-retical idea of the convergency, as it were, in time, of certain groups of forms towards a point of a greater intensity, which he designates by the term " *Polarity ;* " a theory which it would be impossible here to discuss, but which, from its important bear-ings, as well as the deep interest it carries with it as being the last speculation he lived to propose, will doubtless command the closest attention of phi-losophical geologists.

His theory of polarity.

Speaking of the *tertiary* formations, Sir C. Lyell observes, " There are usually so many species in common to the groups which stand next in succession, as to show that there is no great chasm, no signs of a crisis, when one class of organic beings was anni-hilated to give place suddenly to another. This analogy, therefore, derived from a period of the

Continuity in the ter-tiary period.

earth's history which can best be compared with the present state of things, and more thoroughly investigated than any other, leads us to the conclusion that the *extinction* and CREATION of species has been, and is, the result of a slow and gradual change in the organic world."*

Again, he argues at length, from the actual causes which determine the conditions of successive deposits, that, assuming, as we thus must do, the "fluctuations in the animate world to be brought about by the slow and successive *removal and creation of species*," yet, from the local nature of the formations, we *cannot expect* to find conditions such as shall enable us to trace the "*gradual passage* from one state of organic life to another." †

Lamarck's view of breaks.

Lamarck, indeed, held that there may be some gaps in the series greater than we can attribute to mere want of evidence, or hope to see filled up by future discoveries; yet he conceived that the difficulty might be obviated from the consideration of

---

* Principles of Geology, p. 179. 8th Edition. To the same purport, see also Sir H. De La Beche, "Researches on Theoretical Geology," p. 365. To these testimonies I would add one, of even a more decided character, from the very able anniversary address of Mr. W. J. Hamilton, 1855; but being unwilling to spoil so admirable a passage by abridgment, as it is too long for insertion here, I have given it in the Appendix, No. XII.
† Ib. p. 184.

the counteracting influence of a variety of external causes, which are perpetually interfering with the regular order of succession. If these interfering causes did not exist, we might expect an exact continuity of forms; but by these immensely varied agencies of external and local conditions, the progress of some races may be retarded, and that of others accelerated; so that at length wide breaks of continuity may necessarily appear after a long lapse of time.

In many cases too, it must be recollected, that the apparent interruption is confined to certain classes of animals only, and does not extend to others; chiefly among the higher forms; while in the lower, during the same periods, no such interruption occurs, some of them being persistent through many epochs.

But supposing the existence of such apparent gaps or breaches of continuity granted, and that we failed to explain them by any such theoretical suggestions, although we may not yet have hit upon the true explanation or traced *the particular law* in this case, we are sure that *some law* is really involved even in a seeming infraction of a regular

Appeal to general principle of uniformity and continuity.

^Series, and cannot doubt that future discovery will ultimately disprove or explain the apparent anomaly.*

So strong is the inductive assurance of this, that we may safely allow any such apparent exceptions to await their solution without in the least influencing our opinion of the soundness of the broad principle of the continuity of physical causes: a principle of that truly philosophical character which no apparent exception in detail can subvert, or make really inapplicable or unfruitful.

No real interruption.

No inductive inquirer can bring himself to believe in the existence of any *real hiatus* in the continuity of physical laws in past eras more than in the existing order of things; or to imagine that changes, however seemingly abrupt, can have been brought about except by the gradual agency of some regular causes. On such principles the whole superstructure of rational geology entirely reposes; to deny them in any instance would be to endanger all science.

There is no force in such a merely negative argu-

---

* As an instance I may observe that, while this work was in the press, there was announced the discovery of *mammalian* remains in the Purbeck beds, thus *filling up the hiatus* between the hitherto enigmatical solitary marsupials of the oolite, with the tertiary epochs, — having insectivorous teeth, and associated with masses of fossil insects!—Geol. Quart. Journal., vol. x. 420. 475., and xi. 51.

ment; we cannot doubt that the seemingly disjointed portions of the chain must be *really* as much connected as in the more recent instances, where we can see its continuity, and that some future research will as fully and surely close up the apparent breach, as former discoveries have done others once quite as wide.

Thus enough has probably been said to show how completely fallacious is the inference that in such cases as those referred to, because we find an apparent interruption in the observed series of organic remains, therefore we are to conclude a real interruption in the order and continuity of organic existence.   And still further from all sound reasoning or rational analogy must be the inference that, when we find, in a superior bed, animal remains seemingly disconnected with those in an inferior, the actual origination of those distinct species was, therefore, in any way, of a sudden or peculiar kind, disconnected with the preceding order of things, or the orderly progress of natural causes.

No sudden or inexplicable agency necessary.

Throughout all the immense periods of the primeval earth in its manifold mutations, the researches of the geologist present to our contem-

Continued permanence of physical laws in the inorganic world.

plation *two broad facts* in most remarkable juxta-
position; the invariable *constancy* of the nature
and laws of *inorganic* matter and of the forces
acting on it, under all the revolutions affecting
it, on the one hand, coupled, on the other, with the
perpetual indications of *change and fluctuation* in
the forms and functions of *organised existence;* and
the question arises — Can this fluctuation and change
be otherwise than the result of equally invariable
though unknown laws, applying to the organic
world ?

Thus in the *inorganic* world we trace the same
slow and gradual elevations and depressions of continents which we actually witness going on at
present; the same results of earthquakes and
landslips, the action of volcanoes and glaciers, of
submarine currents, oceanic and fluviatile deposits,
irruptions of water over depressed lands, drainage of
lakes, and a multitude of like events, all happening
in obedience to the same identical mechanical and
hydrostatical laws, in the remotest abysses of past
time, as they do at this day; the same influences
of the seasons, and even variations in them, stamped
in the concentric interior rings of fossil trees.

We find the evidence (so beautifully illustrated by the researches of Sir C. Lyell), even back to some of the earliest epochs, of the existence of the same atmospheric conditions; the rain-drops imprinted on the mud; even the obliquity of its descent, indicative of the force and direction of the wind; the very existence of such drops implying the same action of atmospheric electricity and the laws of cohesion; the power of the sun to dry up the mud implying heat conveyed, as now, in the rays of light, thus preserving the impressions of the footsteps of animals on the wet surface left bare by the sea during a short interval, to be covered over by a fresh light deposit by the returning tide, whose recurrence evinces, by consequence, the same laws of cosmical gravitation.

But the unchangeableness of mechanical laws is always found under continual changes of outward conditions; corresponding to which we trace, through the series of organised life, perpetual and unceasing variations of forms and species, yet carried on with such slowness, that we only perceive it by comparison at immense intervals of time. In like manner, of organised life we find some of the conditions

Continued changes in the organic world.

equally unchanged; the animals and plants of those
remote epochs, like those now existing, subject
to the same general physiological laws of respiration and circulation, digestion and nutrition, loco-
motion and instincts; their eyes and ears adapted
to the same optical and acoustical conditions; their
reproduction generally regulated by the same laws;
and *during comparatively limited periods, and identity of external condition, the same permanence of
species.* But amid all these indications of uniformity, when we come to compare the state of
things after immensely long intervals, we find the
nature of whole tribes has been undergoing metamor-
phoses; not arbitrary or heterogeneous in their character, but often repeated in regular correspondence
with other inorganic changes, according to some uniform plan whose law is not as yet made out; but in
all their changes corresponding strictly to the modifications of one common primitive type according
to recondite laws of analogy.

Introduc-
tion of new
species re-
gular, not
casual.

But however little we know of the laws or causes
of these changes, one thing is perfectly clear, *the
introduction of new species was a regular, not a casual
phenomenon;* it was not one preceding or transcending

the order of nature ; it was a case occurring in the
midst of ordinary operations going on in accordance
with ordinary causes. The introduction of a new
species (however marvellous and inexplicable some
theorists may choose to imagine it) is not a solitary occurrence. It reappears constantly in the lapse
of geological ages. It recurs regularly in connexion
with those changes which determined the peculiar
characters we now distinguish in different forma-
tions. It is part of a series. But a series indicates a Due to regular na-
principle of regularity and law, as much in organic tural causes.
as in inorganic changes. The event is part of a regularly ordained mechanism of the evolution of the
existing world out of former conditions, and as much
subject to regular laws as any changes now taking
place. If the series be regular, its subordinate links
must *each* be so ; the part cannot be less subject
to law than the whole. That new species should be
subject to exactly the same general laws of structure,
growth, nutrition, and all other functions of organic
life, and yet in the single instance of their mode of
birth or origin should constitute exceptions to all
physical law, is an incongruity so preposterous that
no inductive mind can for a moment entertain it. It

must have been as truly subject to pre-arranged laws as any case of ordinary reproduction.

Influence of time.

And since, in any conjecture as to the nature of the causes acting to produce these changes, we must admit that *long duration of time* necessarily enters as an essential element, it is obvious that we can in no way form legitimate inferences respecting those causes from any mere observation of natural operations *which do not require time* for their evolution, or conclude against such changes having occurred, even to a great extent, in those immensely long periods, because we do not see them occurring in a short time under our own eyes, in the brief and momentary periods to which our observations extend. Nor in

Appeal to real causes in conjunction with effects of time.

this view is anything implied adverse to the strict application of the truly philosophical principle of arguing solely from real physical causes for the explanation of geological phenomena. Sir C. Lyell expressly includes *lapse of time* as an element among the conditions which he lays down in that grand maxim, worthy to have occurred in the "Novum Organon"—" *When we are unable to explain the monuments of past changes, it is always more probable that the difficulty arises from our ignorance of all the existing agents, or all their possible effects in an indefinite lapse of time, than that some cause was formerly in operation which has ceased to act."*

In a word, if we acknowledge the right mode of investigating the organic phenomena attending the gradual formation of the earth's crust, as in all other cases, to be solely that which proceeds by the analogy of real physical causes, carried on through countless myriads of ages, not by the agency of imaginary convulsive paroxysms, then, by the same rule, the same principles ought to apply in regard to those more obscure changes of an *organic* kind continually going on, whose nature, indeed, is less understood, but which, therefore, form not less an integrant part in the prescribed and beautifully adjusted economy of nature.

Conclusion.

§ II. — THE EVIDENCE DERIVED FROM PHYSIOLOGY.

Physiology essential to the inquiry. GEOLOGY is essentially dependent on physiology; hence any argument derived from the former science, as bearing on the evidence of " creation" in the organic world, must be in some degree an application of the latter ; as, indeed, is manifest throughout the foregoing remarks. But some questions are involved in the present inquiry, which depend on a more particular reference to points of *pure physiology ;* and to these the present section relates.

Researches of Cuvier. From the researches of Cuvier, the whole science of comparative anatomy received a vast, and at the time unimagined, extension, in its application to the organic remains of the ancient earth (first systematically carried out by Von Buch), and the recognition of extinct species, allied to existing forms, in what were hitherto imagined to be either relics of legendary monsters and antediluvian giants, or else mere "lusus Naturæ" created by her "plastic powers."

The intellectual character of Cuvier was equally marked by high powers of generalisation and by a dislike of theorising, or indulging in speculations, as to the causes of the phenomena observed. Yet he inclined, nevertheless, very strongly to the idea of investigating organised structures on the principle which he termed " conditions of existence," or what has been since called " teleology."

Meanwhile, the rising school of Geoffroy St. Hilaire, in proposing the principle of " unity of composition " as that on which alone a philosophical investigation of organised structures ought to be built, was strongly opposed to the followers of Cuvier, especially in reference to the unphilosophical use of the appeal to final causes, in accordance with what has been already explained * : while, in regard to the relations of species, in some respects they pushed their physiological speculations into the regions of conjecture beyond the boundaries within which demonstrative evidence had as yet been applied.

Views of Geoffroy.

---

* See above, Essay I. §§ ii. and v.

In the sharp discussions which these questions underwent in the Academy of Sciences, it was not surprising that, under the high influence of Cuvier's name, a considerable body should have stood out as antagonists to the doctrines of the newer and transcendental school, and stout maintainers of the *safer* dogmas, both with respect to the " teleological " principle, and in opposition to the novel theory of the relations and modifications of species.

Principle of unity of composition.

If we look at the former question in a purely scientific point of view, it amounts to the inquiry whether, in the actual organisation of animals, the " governing principle," or general law, is to be regarded as that of an archetype or common plan, to some modification of which every observed form may be reduced,—or whether we should rather look to the *conditions* under which we suppose each animal destined to exist, and interpret the different structures in their imagined relations to that end. Professor Owen* has justly observed that the two principles of " unity of plan " and " final causes," are " wrongly regarded as antithetical;" and on

* On Limbs, p. 34.

general grounds it must be apparent that these two principles can hardly, with propriety, be put in opposition to each other, or even be classed together : the latter is, in its nature, a more particular and restricted kind of *practical* view of the matter, doubtless of some value in particular cases; while the former is of a comprehensive speculative character, fitted to form the foundation of a philosophical system, which the other never can be, as Bacon has so forcibly pointed out.[*]

If more particular arguments were wanted, Professor Owen has shown precisely, from instances, that the mere investigation of the *uses* of organs continually finds a check in the observation of many cases where organs are introduced whose function or purpose is not fulfilled ; and the more anatomical investigation has extended its bounds, the more clearly have such proofs been displayed, evincing that this principle is an insufficient guide. {Narrow views from teleology.}

Thus, the complex form of a limb, as to number and relative position of the bones, required by the law of conformity to the type, is strictly preserved {Instances.}

---

[*] See above, Essay I. § v. p. 146.   And the whole passage in De Augm. lib. iii. c. 4.

in cases where it is not needed; as is seen in comparing the expanded human hand, where every finger and joint is essential, with the " trowel " of the mole, the " paddle " of the whale, or " hoof " of the elephant; where every bone is equally present, but separately useless from being all enclosed in one case. So, again, the abortive teeth of the young whale are of no use except to prove its relationship with terrestrial mammalia.   Unity of plan is adhered to in other cases where only one or two parts are developed, and the rest are merely rudimentary, or even altogether deficient; but no new part or structure is added.   Nothing is made in vain if it be only made to preserve unity of system.   The view of design has been contracted by the adoption of the false analogy of machines, in which unity of plan is not an object.   The attainment of an end by apparently circuitous means for the sake of obedience to the law of unity is, in fact, the highest indication of design; special adaptation is but a secondary branch of such evidence; and it is only the more striking when brought about in conformity with this higher and governing principle of all animated nature.

With regard to the principle of " unity of composition" itself, some general and vague notions of the uniformity of plan pervading animal structures had been thrown out even by Aristotle; in modern times, Bacon had recommended an inquiry into the *causes* of the diversity of organised forms, and Leibnitz and Newton had both hinted at the idea of a common plan pervading animal structures. These rudimentary conceptions were, perhaps, first developed in a more systematic, yet hypothetical, form in the speculations of Goethe and Oken, and more extensively in those of Geoffroy*, in whose school it had been fully recognised as the most material point of comparative anatomy to establish the analogy of the several functional parts in different species on what was termed the doctrine of "homologies."

Enlarged ideas of unity of composition.

Numerous and striking instances had been long since pointed out, which show, under evident *dissimilarity*, the extent to which real *analogy* is preserved. In many instances the fully developed organ in one animal structure is only found in a rudimentary condition in another; so much so, some-

---

* Principes de Philosophie Zoologique, 1830.

C C

times, as to require the minutest examination to
detect it, yet, still strictly *homologous*, or preserving
the same *relation* to the general structure; and so
strong was the conviction of this general law, that
many physiologists did not hesitate to speak positively of such analogies on the strength of conjecture, where actual examination had not yet detected
them, thus laying themselves open to the attacks of
more matter-of-fact inquirers.

Theoretical
speculations
often useful.
Some of these views, especially those of Oken,
were, perhaps, of too metaphysical a cast to be
usefully recognised by physiologists. Yet, they at
least fulfilled the important purpose of supplying
hints and presumptive conjectures for a more exact
induction to work upon; and whatever, on various
grounds, may have been the prejudices against the
views of the transcendental school, physiologists are
now beginning to pay them the homage of carrying
out and establishing on demonstrative evidence (at
least in regard to this great principle) the ideas which
they suggested.

Analogies
of vertebrate
forms.
To proceed to a more particular view. Goethe
and Oken had thrown out the singular analogical
idea, that the bones of the skull are all vertebræ;

and others had imagined some vague resemblances between limbs and ribs, as connected with the vertebral structure. But it was reserved for Owen to give the full elucidation and establishment of these views, and to supply a detailed demonstration of a principle so essential in the theory of unity of composition of animal forms.

By a close anatomical investigation of the true nature of a vertebra, he confirms the idea of the vertebral character of the bones of the cranium, and includes the whole structure of the vertebral column in a single analogy ; and thus traces the limbs to the development of certain appendages which he has shown to belong to all vertebræ ; but in different cases more or less detached and displaced from those vertebræ, and developed in different degrees, in adaptation to the respective forms and functions.

Owen's researches on the vertebrate form.

Thus the whole skeleton is referrible to one simple scheme or archetype, most resembling the *fish* form ; to analogy with which all the most varied modifications may still be traced.*

---

* These views are most luminously set forth in detail in Professor Owen's small volume " On the Nature of Limbs." London, 1849.

In the Appendix No. VI. I have inserted a more copious abstract, which has received the benefit of Professor Owen's own revision and remarks.

Generalisation of unity of composition.

The idea of "unity of composition," as at first proposed by Geoffroy, like many other ideas struck out by great master minds, was a kind of philosophical prophecy : he did not himself carry out the investigation of it in all its details by demonstrative proofs ; and even in some instances it has been pointed out that he fell into mistakes in particular points of its application. The research remained to be fully followed out by others: and as each of the great divisions of animal life — the Vertebrata, Articulata, Mollusca, and Radiata — had been shown to have *separately a plan and an archetype of its own*, the question which then arose was, Can these four great rudimentary plans be shown to have a yet more comprehensive relation? Can they be included under any *one common and yet more elevated generalisation?*

According to the system of comparing *structure* alone, it was impossible to establish such a point; a different method was necessary : and Von Bär was the first to suggest the principle of studying and comparing, not merely the *adult structure*, but the earlier process of *development* of each form : and in following out this line of research he was able to

From comparing fœtal structure.

indicate characters assumed at successive early stages by the different organs; and thus to trace the relations of species and classes in a way which the examination of the mature structure alone would not have admitted; and he thus showed that, though the *common plans* of the *adult* forms of the great classes are *dissimilar,* yet in their respective developments there is in each *a period* during which an *exact community of plan* prevails: beyond that stage they diverge according to laws peculiar to each.

It has been in carrying out this " developmental method " of comparison that the labours of subsequent inquirers in this field have succeeded in the full establishment, in detailed anatomical examination, of the great idea of unity of composition: and what was at first little more than a philosophical romance, has in their hands risen to the rank of a demonstrated science.

The investigations of Professor Owen forcibly elucidate, not only the correspondence traceable between the perfect organs and functions in different species, but also the relations existing between the *permanent* organisation of the lower classes of animals,

Analogy of vertebrate and invertebrate classes.

and the *transitory* embryonic steps through which
the higher pass, and evince the extent to which the
resemblances expressed by the term "Unity of
Organisation" may be traced between the higher
and lower organised animals; and that it bears an
inverse ratio to their approximation to maturity.
"All animals," he observes, "resemble each other
at the earliest period of their development;" and
he traces out with precision the characteristics
which mark each stage of development as compared with those which permanently belong to
different inferior classes, from "the monad," with
which alone "the potential germ of the mammal
can be compared," up to the vertebrated form in its
different modifications.*

In this way the Annulosa have been analysed to
a common original rudiment of form with the
Vertebrata, by the labours of Savigny, Andouin,
Milne-Edwards, and Newport; the Mollusca and
Radiata, more recently examined, have been reduced to a similar conformity to the same principle,
especially by the labours of Mr. Huxley, who has

---

* Lectures on Invertebrate Animals, &c., 1843.

indicated the archetypal form, belonging to them in common with the former, up to a certain stage, beyond which their peculiarity of character is super-induced.*

In advancing from these researches on *what is*, to their application to *what has been*, and with a view to the discussions which have arisen on such questions, it becomes necessary, in the first instance, to advert briefly to some considerations as to the nature and distinctions of *species* in general. _{Nature of species in general.}

According to the distinction usually maintained in natural history, a *species* is not merely the logical subdivision of a *genus*, but implies the idea of distinctive characteristics *derived from a parent* and the reproduction of like individuals : it involves, not only the consideration of *type*, but of *descent*.

Again, it is within the bounds of *observation of the existing order of things* to recognise the fact that these characteristics in any one species are not *absolutely* fixed, but admit of a certain and often very considerable variation from one individual to _{Varieties.}

---

* To Mr. Huxley I am indebted for a valuable original sketch of these investigations, from which, by his permission, I am enabled to present to the reader a very copious extract in the Appendix No. VII.

another. There are, also, cases in which certain
deviations from the original type are more marked,
and continue to affect several or even many generations constituting " *a variety.*"

These varieties, after a longer or shorter period, *in
many cases* cease to be continued; and it is probable
that the same external conditions, which are favourable to the original type, are less so to the variety,
which is thus more easily checked in its increase, or
at length extinguished. But it is a subject on which
nothing is known as to the real causes which may
give rise to such changes, and on which, therefore, it
is clearly *unwarrantable to dogmatise,* or to reason
upon such failure as if it were a *necessary* law.

Occasion-
ally perma-
nent.

There are also cases in which varieties have been
found to continue so long, and to maintain so completely distinct a character, that it has become difficult, if not impossible, to determine whether they
do not constitute a *sub-species.* And so far as any
speculation can be carried, on a subject so little understood, it would seem most probable that wherever
such permanency has been attained, there has existed
some peculiarity in external conditions which has
been the determining cause for the perpetuation of

such variety, just as some other external conditions of an opposite kind are in other cases unfavourable, and cause its declension or extinction. It is therefore a fair inference, that if the favourable conditions were continued, and the variety were locally isolated from the rest of the species, it would become a permanent type or species.

Within certain limits we observe species *fixed* at the present day: we have, in some instances, proofs from historical monuments and preserved remains that they have not altered within very high limits of antiquity. Some writers refer, for example, to the mummies of Egypt, reaching back to an interval of 3000 years (or, indeed, as much longer as their chronology may dictate); but, in fact, we can go much higher, since we have undoubted evidence of *some* existing species having remained permanent during the countless ages since the tertiary deposits up to the present time.

Argument for permanence of species.

This, however, proves little, since the point to be explained is, that associated with these are found other older and extinct species closely allied to, but different from those which now exist; and the question is as to the relation between the existing species and

Permanence of certain species accompanied by changes in others.

its ancient allied form, or generally between any species existing in one formation or period, and the one analogous, or nearest allied to it, in the next earlier formation in which it had no place.

Difficulty of marking specific distinctions.

Again, it is the fact that there are at present numerous instances in which naturalists are at issue as to what constitutes a *variety* or a *species*; and the very difficulty of determining whether any given character is or is not permanent, is alone sufficient to show how little we ought to press any such principle as the basis of a universal conclusion — still less as partaking of the character of a great and necessary law of nature, on which we can satisfactorily reason from the present to the past, or infer the perpetuity of distinctions so little settled.

The distinctions admitted as those of *species* are indeed in some cases much more minute than those which, in other instances, are considered to mark only *varieties*. An American butterfly (the Vanesso Atalanta) is held to be a distinct species, yet its characteristic is only a single spot on the wing ; nevertheless the entire black skin of the Negro and the white of the European mark only varieties.

In a report formally put forth by the British

Association, it is expressly set down as a DESIDERA-
TUM to discover some sure mode of *distinguishing
real species from local varieties.* "One school," the
reporter observes, "attribute all specific distinctions
to the influence of external agents ; another regard
the most trivial circumstances as fixed;" but he
recommends a just mean as the preferable view.*

Professor Pictet also enlarges on the difficulty of
distinguishing species, and gives instances of such
ambiguity of character.†

On this subject I will again cite the authority of
Dr. Carpenter: — " The uncertainty of the limits of
species is daily becoming more and more evident ;
and every naturalist is aware that a very large
number of races are usually considered as having
a distinct origin, when they are nothing more than
permanent varieties of a common stock."‡ And he
then goes on to point out the course which the true
naturalist must take in the endeavour to define them
more precisely, by attending not merely to form and
structure, but to " the *whole* natural history of a

---

* British Association Report, 1844, p. 219., On Ornithology, by the
late H. E. Strickland, Esq.
† Geological Quarterly Journal, No. v. p. 48.
‡ Principles of Physiology, Art. 542.

reputed species," to ascertain and limit its real character.   But, it may be fairly remarked, if such
" permanent varieties." present characters so constant,
the real bearing of the question is evident: Why may
not other or all " reputed species " have once originated in the same way ?

Increase in
number of
known
species.

Among the various considerations necessary to be
taken into account in forming a fair judgment on the
whole question, another somewhat material inquiry
arises : as discoveries and explorations extend and
increase, so must the *number* of known species
eventually extend and increase in an almost incalculable ratio ; and the inference clearly is, that as
new species are thus continually being *inserted* between other allied species already known, it is
evident that the specific differences between each
must tend to diminish continually, and *all species*
tend to be connected by more and more close
affinities.

Real num-
ber of
species in-
finite,

It has been stated, on good authority, that by the
recent progress of research the number of *known*
species of plants and animals has been *doubled* in
some classes, and *quadrupled* in others, within the
memory of persons now living : and, considering the

number of *new species* constantly reported in every fresh exploration, in every successive number of every journal devoted to natural science, we cannot but suppose the increase to continue at least at an equally rapid rate.

But while the number of species thus tends to become infinitely great, the extreme difference between man (let us suppose) at one end, and a zoophyte at the other end of the scale, is constant and *finite;* hence the average difference between any two species tends to become *infinitely small.* Multiplied by the number of species, it must still be equal to a *finite* quantity; and the product being *finite,* if the first factor be *infinity,* the second must be *zero.** {And differences evanescent,}

The close approximation in character between many allied species has led some philosophers to speculate on the real difficulty of any absolute and philosophic distinction between them; for all *prac-* {Yet sufficient practical distinction.}

---

* Mr. W. J. Hamilton (in his address to the Geological Society, 1856) has represented this passage as a fallacious argument in support of transmutation.

If I had brought it forward *as such,* it would no doubt be chargeable with that accusation. But I think a reference to what precedes will at once show that I *do not* adduce it *as an argument in support of transmutation.* The remark refers entirely to *existing* species; and it is only brought forward as one of the general considerations necessary to the discussion of the entire question. See also Appendix No. X.

tical purposes, indeed, and in reference to the *existing state* of knowledge, and any state which it may be likely for a long time to assume, there is in general sufficient ground of distinction.

The alarm felt lest the power of making specific distinctions should be thus done away, and with it all substantial science of classification should disappear, is seen to be groundless when we observe

Species permanent for the present era.

that it is on all hands allowed that species are likely to be as strictly permanent as at present for many thousands, perhaps millions of years to come, *provided the external conditions remain the same.* When the distribution of continents and oceans, the elevations of land, the direction of currents, and the like circumstances, shall have undergone a great and notable change, influencing the climate and productions of existing lands, and even presenting new regions for the diffusion of life, we might then well expect that some existing forms might be lost, and that such a gradual change of species, and eventually even of whole genera, might at length take place as would fully exemplify, and account for, the observed changes in ancient formations.

Races of men, whether

Much discussion (as is well known) has arisen on the question whether the *different races of men* are *varieties* of one species, or *distinct species :* and it species or varieties. seems to be at present the prevailing opinion that they are *varieties* merely.* But the question *how*, by what steps or processes, did such large and fundamental differences arise? entails more important consequences than many, in their zeal to maintain a single origin, seem to perceive. It is clear that these differences are fully as great as those which in many other cases are allowed to constitute distinct species.

If in the case of man they have occurred as transi- Difficulties of the question. tional varieties, how comes it that they have become so inveterately permanent? And if those changes have all occurred within the lapse of a few thousand years of the received chronology, it cannot with any reason be denied that similar changes might occur among inferior animals, and become just as permanent. And if so, changes to an indefinitely greater extent might occur in indefinite lapse of time. If these changes take place by the gradual

---

* For example, Dr. Pickering, after an extensive investigation, concludes that there are either *eleven* species, or only *one*, of the human race. But upon further examination he decides in favour of *one*, and thinks the original seat of man was in Africa. (The Races of Men, &c., by C. Pickering, M D. Lond. 1850.)

Others have assigned *six*, or other numbers of species: the author of the "Vestiges" supposes *two* local centres necessary. (P. 220. 6th Ed.)

operation of natural causes, it would be preposterous
to deny the possibility of equal or greater changes
by equally natural causes in other species in equal
or greater periods of time. The advocates of the
fixity of species would argue that the single spot on
a butterfly's wing, which constitutes a species, never
has changed, and never can change, without a
miracle; and yet the vast differences between a
European and a Negro or Australian are mere
modifications of one parent stock by natural causes
in the lapse of a few thousand years!

The peculiar characters of the Negro race are
recorded, as prominently marked as at present, in
the ancient Egyptian paintings, which may go back
3000 years or more.* Here, then, is a *variety* which
has been *permanent* for at least that long period;
a period, too, which has been expressly relied on
by many to prove the permanence of *species* by
appeal to these very monuments. And then we
have to ask, How long must it have taken, at this
rate of imperceptible progress, to have been developed out of the original stock?

---

* See Kenrick on Primeval History, p. 20.

Another instance has been much dwelt upon, the Varieties of the dog. so called " varieties " of the dog, presumed to be derived from a common stock ; but how long since, is undetermined. Yet in these varieties (in which even the form of the cranium greatly differs) it would be difficult to deny that the distinctive cha-racters are permanent, at least in some of the more marked instances, and under the continuance of the same external conditions ; and that each race, when preserved isolated under such conditions, would remain permanently distinct.

Much stress has also been laid by some on the Hybrids. asserted sterility of hybrids ; though, in truth, it affects very little the general question ; while its very limited evidence dependent only on a few isolated facts, occurring in a state of domestication, is utterly insufficient for the foundation of any general law. The cases commonly referred to should be regarded by an unprejudiced mind as probably exceptional, under peculiar conditions, and not to be dogmatised upon, as involving any real and necessary law of organised existence. As there are limits beyond which union will not take place, so within these there may very probably be *certain* *limits* of still nearer affinity, *beyond which* sterility in the offspring prevails, but which have not yet been determined. The recurrence to the original type often observed, only proves that conditions are not favourable to the continuance of the variety.*

Conclusions of this kind empirical. And of the very positive assertions so liberally made in these and the like cases, it is to be observed that they are, at best, merely *empirical* conclusions, wholly unsupported by any wide analogies, or explained by any known causes which can confer on them the character of real natural principles.

Hypotheses of immutability of species. Yet the immutability of species, as something essential to their nature and inherent in it, has been upheld by a large section of naturalists — and still more strenuously by some who are not naturalists — in this country, with a degree of positiveness and even vehemence, which the mere *negative* character of the evidence could never justify, and which it would be difficult to account for, so far as any arguments of a philosophical nature may be supposed to influence the opinion.

It is indeed difficult to say what extent of mysticism is not connected in the minds of some with

* On the subject of hybrids, some important remarks will be found in Dr. Carpenter's Physiology, Art. 545.

the notion of the immutability of species. Even
such sober naturalists as MM. Agassiz and Gould *
speak of it as dependent on an " immaterial principle " essential to animal life.

But in other schools, especially on the Continent, Opposite
opposite views are extensively maintained, and pro- views.
bably gaining ground. In the case of plants more
particularly, it is simply as a *question of facts*, that
some eminent botanists view the matter. Thus one
of the most distinguished foreign naturalists, Prof.
Schleiden of Jena, after giving a variety of illustra-
tive instances, thus sums up the state of the case : —

" We know that varieties once formed, when they
have continued to vegetate under the same conditions for several generations, pass into sub-species ;
that is, into varieties which may be propagated with
certainty by their seeds. How, then, if the same
influences which have called forth an aberration
from the original form of the plant, continue to act
in the same way, not for centuries or tens of centuries, but for ten or a hundred thousand years,
will not at last, as the variety thus becomes a sub-

* Principles of Zoology, p. 43.

species, so also, this, become so permanent, that we shall and must describe it as a species." *

Even in this country, Dr. Lindley has ventured to state not less alarming facts respecting the class Thallogens (including the seaweeds, fungi, and lichens), to the effect that "in their simplest forms all trace of series is missing," and that their species seem all convertible into each other under particular conditions.†

On these questions I have much satisfaction in referring to the opinions of Dr. Carpenter‡, put forth with equal candour and freedom from prejudice, and with a union of caution and enlargement of view which eminently commends them to the convictions of the reader.

He observes§, "Our belief that the new beings formed by the process of reproduction always closely

---

* "The Plant," &c. by M. S. Schleiden, M. D., Professor of Botany, Jena.   Transl. London. 1848.   Lect. xi. pp. 272. 290.
† Vegetable Kingdom, 1846, p. 5.
‡ A number of important examples are collected by Dr. Carpenter, "Physiology," Art. 547, 548.
  I do not here profess to go into details of particular instances, but many such, of transmigrations of existing species, which have been collected, are unassailable evidence as far as they go.   (See for example, "Vestiges," p. 136. *et seq.* 6th Ed.)   See also a paper in the British Association Report, 1852, sectional proceedings, p. 68., by Major Munro, F. L. S., who limits transmutation to plants of the same genus.
§ Physiology, Art. 517.

resemble the parent stock, is certainly founded upon a limited induction from observations made upon the higher classes of plants and animals. Reasons have already been given for the opinion that the same germ may assume very dissimilar forms according to the circumstances under which it is developed: and knowing, as we do, how readily the simpler classes of organised beings are affected by changes in their external conditions, it is not difficult to admit the possibility of their forms being thus greatly modified, as well as of the continued propagation of the varieties thus produced."

As to the origin of varieties, it has been admitted by most physiologists as a general principle, that those peculiar "vital forces" (of whatever nature they may intrinsically be) are always the main acting cause, but are subject to modifications from external causes: and hence it is the preservation or interruption of the balance between these antagonist causes which determines the formation or the modification of the type, so as in the one case to keep it up, in the other to produce *varieties*.

Origin of varieties.

Again, anatomical inquirers have chiefly confined their investigations to the normal forms of organised

beings. Autenrieth of Tübingen seems to have been
one of the first to dwell upon the necessity of taking
into account, in comparative anatomy, not merely
the *perfect* adult structure, but all the varieties of
abnormal structure which are occasionally witnessed,
and to which the absurd name of "lusus naturæ"
has been given, just as organic fossils were once ascribed to the plastic powers of nature, or elevations
and subsidences to mysterious convulsions. He observes, "These varieties . . . are not haphazard formations; they are the remains of structure common
to all embryos; they indicate the transitions through
which man and all other animals are passing from
their embryonic condition to the adult." Should
anything interfere with this transition, the forms
persist; and this constitutes what is improperly
called *a variety*, and supposed to be something
deviating from the regular law. "But the laws
of deformation are as regular as the laws of formation. The varieties are arrests of development; they
prove the unity of organisation and of type with
which Nature starts in the formation of all that
lives." *

- * Quoted by Dr. Knox, "Great Anatomists," &c. p. 59., who also cites
many instances of imperfect development, and other cases bearing on the
question, p. 108.

Such considerations form a necessary preliminary to any examination of the theories of change of species in the earlier epochs of our globe. Cuvier (as it has been well expressed)* was the first to give " a history of the earth, not founded on fables, but on facts ; " but it was incorrectly called a " theory of the earth ; " it was really only a " history," whereon to build a " theory."

Causes of changes in species in past ages.

On the theory — the philosophy of that history — it was reserved for others to speculate. The laws which regulated the succession of living forms in the different epochs of the earth's existence, and those higher generalisations which might tend to indicate the physical causes of those changes, were the objects of inquiry to a considerable section of continental physiologists. When philosophers began to speculate on the possible causes of changes of species in the ancient world, it could hardly be otherwise, in an inquiry of so wide and novel a character, than that the several hypotheses started should be but imperfect in detail ;—should be rather guesses at and approaches towards the truth ; and

---

* Knox, p. 25.

should, even in the opinion of adherents to the general principle of some regular process of evolution, be seen to require revisions and retractations in minor particulars.

Cuvier, while he professed a rejection of all hypothetical speculation, appears to have been strongly prepossessed with one hypothesis—that of the essential and eternal *immutability of species.* At any rate, under the sanction of his name it has been since maintained by many of his followers with a degree of positiveness not easy to account for on any merely philosophical grounds.*

Theory of transmutation.

But an opposite opinion began to be taken up by those of a different school, intimately connected with the more speculative and transcendental views already alluded to.

Connected with the principle of unity of composition.

When Geoffroy and his coadjutors were engaged in upholding the unity of composition of all animated structures, it was regarded by them as a natural consequence that, as all the details which mark dif-

---

* Dr. Knox, however, the personal friend of Cuvier, states that, at least latterly, he was much inclined to modify his opinions on this point, and quotes one passage in which he says, " Nous ne croyons pas même à la possibilité d'une apparition successive des formes diverses."—Great Anat. &c. p. 44., also p. 27. The expression, however, is perhaps ambiguous.

ferent species were the mere modifications of greater or less development of the different parts of the primary plan common to them all, so those developed parts had no essential or permanent place in the nature of the species. Hence, naturally arose the further idea of the possible migration or transition from one species to another, or rather from one species to a *new* modification.

Thus the theory of "unity of composition" was, in the minds of many, closely allied to that of " transmutation," which seemed to be a sort of natural sequel to it, and was, among a large school of the continental naturalists, associated with the advancing prospects thus held out to their view of the system of nature, and the attainment of a more transcendental theory of her operations.

From the principle of unity of composition as applied to the existing animal world, it was by a natural extension that this school of physiologists were led to infer that there had been an equally close analogy preserved in *time*, and that there had been a continuous succession of the several species of the animal world, of which only a few detached fragments are preserved to us, as disclosed by geolo-

Applied to past changes.

gical research : a succession, as they contended,
which took place in the natural course of reproduction by continual small deviations from a primitive
type, according to the influence of external conditions which varied from one epoch to another,
yet were determined alike by fixed and regulated
laws, from the present era backwards into the
abysses of past time.   They conceived that in none
of the varied forms of organisation which we trace
has there been any new fundamental principle
introduced; all are but modifications of forms now
existing.   Thus throughout all nature, present and
past, external forms are mere accidents ; development
of parts in excess or in defect as changing causes led
to such necessity, and modifications of parts according to the functions required to be exercised
under the particular conditions.   The fundamental
unity of principle is that which alone is permanent and invariable, and which admits of endless
adaptations according to the varying conditions of
existence.

Theories of
develop-
ment.

The speculations of Lamarck were founded, in the
first instance, on observation of the fact of the near
approximation of different species.   He contends

that the farther our researches extend the greater we find the difficulty of distinguishing species. Apparent interruptions in the continuity of the series are continually being filled up by new discoveries, and what were supposed broad lines of separation effaced. " Everything," he observes, " passes by indivisible shades into something else."

De Maillet (under the anagram of Telliamed) and Lamarck were perhaps those who followed out speculations on the transmutation of species to the greatest extent. Not contenting themselves with asserting cautiously the philosophical grounds on which the close unity of all organised structure is supported, they pursued their hypothesis into details of the most minute, and often most extraordinary, kind. Lamarck's theory, in particular, in truth presents many salient points easily open to attack, and some which are readily susceptible of being held up to ridicule. His principle, that organisation is the result of function, not function of organisation, seems, at first sight, somewhat like a mystified version of that very doctrine of final causes to which many of his school so strongly object. At any rate his theory was carried out to such unwar-

rantable lengths in many instances, as to discredit
even the more sober primary assumptions on which
it was founded, in the eyes of his opponents; while,
as might be expected of so imaginative a speculation,
increasing knowledge of facts has led to its abandon-
ment, at least in its full extent, on the part of many
who yet strenuously uphold the same broader principles.   Thus Dr. Knox*, one of the most zealous
supporters of the principle of transmutation in this
country, speaks very slightingly of Lamarck, and
regards his theory (in its extent and detail, at least)
as of little weight or authority at the present day:
and the necessity of great modifications in the theory
is admitted by the author of the " Vestiges."†

Argument
from ana-
logy.

But in order to be applicable to the facts of the
ancient earth, any process of transmutation must be
imagined to extend, not only to minor features, but
to a total change even of the characters which mark
whole species and genera, so that entire classes and
orders may in the course of indefinite progression
disappear, and be replaced by others of a different
kind.   Lamarck endeavoured to support his theory

---

* Knox, p. 72.   † Vestiges, p. 143. 6th Ed.

by adducing instances actually occurring within our experience, of slight variations in species; but of more extensive changes confessedly none could be adduced from *existing* facts. Yet it was argued in such cases we must bear in mind the necessity for the introduction of another important element — *the influence of time.* Thus the allegation of actual instances of slight changes in finite times is unanswerable as far as it goes; but the absence of such evidence for greater changes in *finite* times is no argument for their non-occurrence in *indefinitely extended* duration.

Influence of time.

As in general, from connecting the conceptions of physical causes with that of immense duration of time, we obtain a very different view of the magnitude of the effects they may produce; so the continental school of transmutation regard all differences in the succession of species of past existence as rather of a chronological than of an essentially physiological kind, due to the lapse of time rather than the introduction of new physiological elements. Thus, De Blainville says that " species mark an epoch in time, not a distinction in animal nature."*

---

* Knox, p. 207.

Opposite views of some geologists and naturalists.

But a large section of geologists and physiologists, including many names of the highest distinction, have been strongly opposed to all speculative theories as to the probable origin of those changes in species which geology discloses. This, in many instances, has been the result simply of a rigid — perhaps over rigid — adherence to the rule of appealing to *facts* only, and allowing of no *hypothesis;* a rule which, however strictly inductive, if carried to an extreme would defeat the grand purpose of induction; and partly from imagining that, because the precise theory of development from lower to higher forms is untenable, therefore all hypotheses of the same kind are inadmissible.

Professor Pictet* denies the transmutation theory, on the ground that to produce such changes the " powers of nature " must be supposed to have been much greater in the earlier periods than we now find them to be; not allowing, apparently, for the indefinitely long time they had to work in; and, in opposition, he is disposed to maintain a series of sudden introductions of new species, which would imply

---

* Geological Quarterly Journal, No. V. pp. 53. 56.

powers greater still : though he afterwards seems not
to consider this as quite satisfactory.

Professor Sedgwick, while he upholds the *progressive* scale of organisation in past epochs, yet earnestly repudiates the development theory, maintaining
that the changes of form " mark a gradual evolution
of creative power, manifested by a gradual ascent
towards a higher type of being. . . . . But the elevation of the fauna of successive periods was not
made by transmutation, but by creative additions." *

But the question is as to the *nature* and *law* of
those additions, and in what particular *sense* or
manner they were " creative ; " whether they were
not made according to some determinate law, and in
some fixed relation to those which preceded them or
were most nearly allied to them. The question is
one demanding calm and patient philosophical analysis, and will be the more fitly and worthily discussed the more expanded and generalised the views
adopted of geological and cosmogonical order and
progress, the more free the inquirer may be from
bias and prepossession of other kinds.

*Real ques-
tion of
creation of
new species.*

---

* See " Studies of Cambridge," 5th Ed. Introduction, pp. 44. 154. 216.

Meaning of
the term.

The term "creation" indeed, especially as respects new species, seems now, by common consent,
to be adopted among geologists as a mere *term of
convenience*, to signify simply the fact of origination
of a particular form of animal or vegetable life,
without implying anything as to the *precise mode* of
such origination — as simply involving the assertion
that a period can be assigned at which that species
appears, and before which we have no evidence of
its appearance. In this sense there can be no objection to its use, but it should be carefully guarded
against possible misapplication.

Dr. Knox observes, " It was the opinion of
Geoffroy that there never had been but one creation: this (he adds) is also my own opinion. I
believe all animals to be descended from primitive
forms of life forming an integral part of the globe
itself, and that the successive varieties of animals
and plants which the dissection of the strata of the
earth clearly sets forth are due to the occurrence of
geological epochs, of the power of which we cannot
form any true conception." * This might seem like

---

* Knox, p. 109.

a leaning to the convulsionist theory, were it not more clearly explained in the following passage : —

" We know not, then, the cause of the specific and generic differences in animals, nor why such differences continue fixed for a period — the historic period, for example; they depend, no doubt, on secondary laws, which some future Newton may discover." *

The subject of Professor Owen's investigation, before referred to, acquires a higher interest when viewed in reference to the question of the progression of animal forms in past epochs of our globe ; and to this view he has pointed † as appearing to assign a certain conformity in the order of ancient existence with that of development of the archetype, as indicated by these anatomical researches. Professor Owen, however, is specially desirous to be understood as applying his conclusions solely to the *order* and *law* of succession, without any attempt to assign a *cause or to trace its origin.* But, though anxiously disclaiming the charge of being supposed in the slightest degree to support

Change of forms by regular laws.

---

* Knox, p. 109.   † On Limbs, p. 86.

E E

the development theory, he nevertheless expresses
himself in the most guarded and strictly philosophical language in reference to the possible modes
of explaining past changes of species, and proposes no
opposite hypothesis of " creations " : — " To   what
actual or secondary causes the orderly succession
and progression of such organic phenomena may
have been committed we are as yet ignorant." *   I
would only venture to add, that it must have been
committed to *some* regularly-ordained causes is surely
the verdict of all inductive philosophy.

Sir C. Lyell, though strongly opposing Lamarck's
theory of development, yet freely admits, in a geological point of view, that " *If* the doctrine of
changes operating in an indefinite lapse of time *be
tenable,* we are at once presented with a principle of
incessant change in the organic world, and no degree
of dissimilarity in the plants and animals which may
formerly have existed, and are found fossil, would
entitle us to conclude that they may not have been
the prototypes and progenitors of the species now
living." †

---

* On Limbs, p. 86.   † Principles of Geology, p. 545.

Though the precise idea of development in one simple series, from the lowest to the highest forms, be clearly untenable, yet in what sense *some regular evolution* of successive forms may be admissible has been here discussed, and with a sufficient result, if it be only allowed to be so far conceivable as that no sudden interruptions of natural order are necessary to be resorted to in order to explain the phenomena.

Nor is the question of non-progression without a bearing on this point; since the narrower the limits of real variation of species in different epochs, the less difficult would be the application of any theory to account for them.

We may here properly advert to another point closely related to the question of the origin of species — the consideration of the *different faunas and floras characterising different districts of the earth.* These, as is well known, are in many instances strikingly distinct, even in regions situated near each other, and as remarkably similar in some other regions separated by intervening seas. In certain cases, too, we have a singular *parallel and corresponding series* of species in two different regions

Local faunas and floras.

or continents, though no species are the *same* in
both.

Hence naturalists have come to the conclusion
that these respective systems of species must have
been thus originally constituted by distinct origination peculiar to their respective localities. Thus,
MM. Agassiz and Gould observe, — "There is only
one way to account for the distribution of animals
as we find them; namely, to suppose that they are
autochthonoi — that is to say, originated like plants
on the soil where they are found." *

Theory of
specific
centres.

But this view has been carried out in a more
precise form by the late Professor E. Forbes, in his
theory of " Specific Centres," or " Centres of Creation." This is founded on the general fact, authen-
ticated by accumulated comparative observations
(after allowing for some apparent exceptions), that
numerous regions and districts of the earth are well
marked out, each characterised by a flora and a
fauna on the whole peculiar to itself. A given
species (for example) is found to be peculiar to a
certain district; the numbers of its individuals di-

---

* Principles of Zoology, p. 179.

minish as we ascend to a particular point of geological time (subsequent, of course, to the eocene period); they also diminish at greater distances round a particular locality. Hence we recognise a single point of origin, both in space and in time, for that species: this is termed a " specific centre."

Again, in many instances, such points of origin for many distinct species occur near together, in local position, in the district characterised by those species; hence it is argued that at such points those species must have originated from an individual or pair (as the case may be), respectively the " prototype" or "protoplast" of each such species, from which that species, marked by constant specific characters, has been derived in the ordinary course of propagation. It is, however, distinctly admitted by the author, that *in what way* that prototype was formed or produced *we know not:* hence, then, and in this sense, such spots are termed " centres of creation."

And as this applies to existing species, so, the author argues, by analogy (as all organised life is part of a uniform connected system), we may fairly suppose it true of extinct faunas and floras also.*

Bearing on develop-ment.

As to the bearing of these speculations on the question of development or transmutation, it is to be observed that the only point of the theory of specific centres which bears at all on the question .is the descent of local species from a prototype of the .same identical specific characters. The author ex-pressly says, — "We have no experience of the individuals of any species being produced otherwise than from individuals of its own kind " — which is, in fact, simply *assuming* the whole question at issue. But if the *descent* of the species, unaltered through all time, were granted, the question would still remain open as to the first origin of the " protoplast," whether one or many.

It is, however, contended that distinct centres of origin are inconsistent with that view of develop-ment which traces all species to one common origin. But I am unable to see in what way it would follow

---

* Of this theory various details have been published by its lamented author : but I am happy to be able to present the reader with a con-densed abstract of its leading principles from his own pen, with which he kindly furnished me, and which will now acquire a peculiar value and interest. (See Appendix, No. viii.)

that the earliest forms of a given existing species, in
a particular locality, were not modified forms of
other species previously existing there; or on what
grounds it could be inferred that these, in their turn,
had not been modified from earlier forms; or these
again, in still more remote epochs, derived from one
common source.

Thus, on the whole, the theory of " specific
centres of creation " really bears very little on the
question of development; it merely shows that
whatever reasoning (on that theory) may be applied
to the origin of species in general, must be restricted
to their origination in or near *a certain locality*; in
all other respects the argument, whether for or
against transmutation, remains just where it stood,
without reference to the theory of centres.

Still less do these facts of corresponding series of
species in different continents or large districts of the
earth affect the question of transmutation, since within each district such succession might have gone on
separately from two parallel species now extinct. It
is even possible that the more remote progenitor-
species in the two districts might have been more

Merely as to locality.

nearly allied, the divergency increasing with each successive step.

Conclusion. Correct statement of the general facts.

On the whole, then, comparing the limited extent and purely empirical nature of our knowledge of *species* in the existing state of things, with the positive evidence of past changes, it would seem that the more correct statement of the general fact would be simply that *species (within certain limits of deviation) are permanent during very long periods, but beyond those periods a change, in some sense, occurs,* and this bears some relation to changes of external conditions. But under the same change of conditions one species may be highly susceptible of, and sensitive to, the influence of that change, while another may be insensible to it. Thus one may remain permanent, while another may undergo change, or be exterminated.

The real alternative.

The only question is as to the sense in which such *change of species* is to be understood,— whether individuals, naturally produced from parents, were modified by successive variations of parts, in any stage of early growth or rudimental development, until, in one or more generations, the whole species became in fact a different one; or whether we are to believe that the *whole race* perished without reproducing itself, while, even during its continuance, independent of it, *another new race,* or other new individuals (by whatever means), came into existence, of a nature closely allied to the last, and differing often by the slightest shades, yet *unconnected with them by descent;* whether there was a continuation and propagation of the *same principle of vitality* (in whatever germ it may be imagined to have been conveyed), or whether a *new* principle or germ originated independently of any preceding, *out of its existing inorganic elements;* to which the principle of vitality (in whatever it may consist) was superadded in some way as yet unknown.

And if it be alleged that even in the newer formations there are some instances in which the new species have little or no analogy or resemblance with the preceding, and therefore cannot be descendants or modifications from them, these are merely cases of those *apparent gaps* or interruptions of series which have been already so much discussed, and which on broad principles we may be assured cannot be any *real* violations of continuity. After all, we cannot by any means pretend to limit modifications of varieties to the narrow boundaries of differences observed in the momentary period to which our observation extends.*

Organisation out of inorganic matter. Speaking of the opinion of the formation of organised beings out of their inorganic elements, Dr. Carpenter observes—" It has been maintained by many philosophers who have regarded all matter as, in some sort, animated; and, although it has been principally urged in reference to the lowest class of beings, it does not seem possible to limit its application *if it be really valid.* . . . . Some naturalists of the present day are disposed to admit this also, and to account for the changes in the races of plants and animals which geological researches reveal."† In a note he adds — " Such a doctrine is impossible to refute otherwise than by an appeal to facts. No

---

* I have inserted a few sentences in this place in order to meet one of the criticisms of Mr. W. J. Hamilton, in his Address to the Geological Society, 1856, p. cxvi. As to what he adds respecting the vital principle, that is altogether a matter of *opinion* with which I do not meddle. All I contend for is the general principle, that *in so far as* it is a matter of *scientific* inquiry, it must evidently be referable to physical causes.

The superaddition of the vital principle to an organised material structure fitted for its reception, may be open to variety of theory as to its origin; but I think the extension of sound analogy can only lead us to regard it as just as much a result of some unknown combinations of regular physical conditions as any other natural phenomenon: if any one wish to contemplate it in any other light, he passes out of the region of *science* into that of *mystery.*

† Art. 516.

such new creations are known to us at the present time; and therefore it can only be argued from *analogy* that they ever existed. We may believe that there exists in all matter a tendency to become organised, without relinquishing the doctrine that for the maintenance of such tendency a previously existing organism is required, to collect and unite the scattered elements by the powers with which it alone is endowed. . . . . That species have in all ages of the globe maintained their present uniformity and narrow limits of variation the author is not disposed to assert; and he thinks that many facts tend to prove the relaxation, at former epochs, of the strictness of the laws which are at present regarded as governing their modification and reproduction."

The temperate and candid tone of these remarks offers a very satisfactory contrast to the one-sided and peremptory dogmatism we too often encounter on this subject. But it may be asked, is any "*relaxation*" in former epochs necessary to be supposed, when we simply take into account the enormous and inconceivable length of time implied in such periods? Do we require anything more than the strict observance of the very same laws of slight

changes of forms accompanying corresponding changes
of condition, acting through periods of incalculable
length?

Sponta-
neous gene-
ration.

What has been termed " spontaneous generation "
has been, as is well known, very generally rejected
by modern physiologists, who adhere to the dogma
"omnia ex ovo." Yet there have not been wanting
some who have advocated the opposite view. Experiments in which the production of some of the
lower forms of animal life have been prevented by
the strict exclusion of the atmosphere, it has been
well argued, are not conclusive, because the presence of the atmosphere may obviously be necessary
in other ways than as transporting ova or seeds.

The case of Entozoa has been philosophically
viewed by some as more probably a case of development from physical conditions*, while the incredible marvels resorted to in supposing an
universal dissemination of the seeds of plants and
ova of animals, tend to throw discredit on the doctrine altogether rather than facilitate the explanation.

But there is a question bearing on the whole

---

* See Art. Zoophyte, Encyclop. Britt., 7th Ed.

inquiry which does not seem to have been sufficiently attended to, viz., what in fact is the original germ or element which constitutes the *essential* principle of a seed or ovum ? It becomes a question of *degree.* The ovum from the parent-stock is *not complete* at first. If it commence as a simple cell, *afterwards* modified and perfected, why may not the same process take place under *other* circumstances ? We do not yet know how elementary is the first rudiment which may develop into a seed, or ovum, or what determines it to become one, or why the process might not go on without the presence of the parent plant or animal.

But while on these points we have confessedly no *positive* evidence, it is fairly open to conjecture, whether the views now universally adopted by the most enlightened physiologists, of the great principle of *the unity of primordial structure* of all species may not be peculiarly suggestive with reference to such questions as those now before us.

Legitimate conjectures from primordial unity of structure.

We have already considered that grand fact on which the whole theory of unity of composition is based, — the existence of a stage in the early evolution of every class and order, during which a community of form belongs to them all. At this stage there exists no difference between them; and out of this primitive common germ or rudiment any one of the more distinct specific forms might, as far as we know, be equally produced, provided the *determining causes* for that particular modification were present. Of the nature of those specific determining causes nothing whatever is at present known. It is therefore clearly impossible to say how far great changes of condition in external agents, or equally great changes as slowly and gradually advancing in the more hidden internal agencies of the animal economy, might not in past ages have operated to determine successive changes in the evolution of germs, originally the same, into great varieties of organisation.

§ III.—GENERAL CONSIDERATIONS ARISING OUT OF THE PRECEDING EVIDENCE.

THE questions to which the preceding observations refer have been the subject of much vague speculation and vehement controversy. But in the present discussion, equally desirous of avoiding the one, and deprecating the other, I wish to take a perfectly unbiassed and dispassionate view of the real tenour of the evidence; and more especially to analyse certain arguments often brought forward, and regarded as based on indisputable principles, which nevertheless appear to me involved in considerable doubt and fallacy: And though, in some instances, they boast the sanction of names eminent in physiology and geology, yet the question is rather one of general principles of reasoning than of precise scientific details; and thus, without pretending to impugn their *science*, I venture to call in question their *logic*.

In the first place, then, the belief in *the essential and inherent immutability of species*, not only in the

Analysis of the reasoning founded on the foregoing evidence.

Argument for immutability of

species from permanence of natural laws.

present state of things, but as an eternal law of Nature, extending backward through all the countless ages of the ancient earth, has been upheld confessedly on the limited experience of modern observation, but thence extended by analogy in the same way (it is alleged) as *in the case of other great natural laws.*

But difference in evidence.

This argument, however, appears to me altogether unfounded. Of the operation of other great natural laws, through all the series of past ages, *we have direct evidence.* The laws of gravitation, heat, light, equilibrium, and the like, present positive proof of their influence in the records preserved to us through all geological time; whereas of the permanence of species in those past epochs, except within certain limits of particular formations, we have *no evidence* whatever : on the contrary, the *apparent* phenomena (to say the least) are all *opposed* to it ; and it is the very question at issue, whether those perpetual *changes in species* which we observe, are to be considered real gradual variations in the development of organisation, or to be explained in any different way.

Mechanical

Again, in regard to the mechanical laws of the

inorganic world, *reasons* can be assigned, and calcu-

lation appealed to ; whereas, in regard to *organic* life,

the conclusions, such as they are, are wholly *empiri-*

cal.

laws proved
by prin-
ciples;
these only
empirical.

The case is by no means analogous (as some seem to suppose) with that of the planetary perturbations. They would argue that, as the law of elliptic motion has held good, subject to those small deviations, through all the past existence of the system, so the permanence of species must have held good, allowing for like small deviations from type in occasional varieties. But the cases are obviously not parallel : in the former, the perturbations are all parts and consequences of the same principle and law of gravitation, perfectly understood and demonstrated by calculation to be in long periods perpetually compensated, so as to preserve the system. In the other case, the law (if it were such), is merely *empirical :* we know of no principle or reason for it. The deviations are no part of it, or consequence from it ; nor can their extent be predicted. We know nothing of any causes acting, nor of any conditions which confer on it the character of necessity.

Again : the utmost extent of proof which can be

adduced in support of a permanence of species

Dependent
on certain
conditions. amounts to this, — that *they are permanent so long as
some peculiar external conditions remain the same.* If
those conditions were materially altered, no present
experience will enable us to predict the result. In
past epochs we know that the *conditions* were repeatedly altered; and we know that certain *changes*
of species accompanied the altered conditions. Now,
the stability even of the planetary system is constant
only as long as the *same conditions* remain. Let
them be altered in the slightest particular, — let a
period, an eccentricity, or an inclination, of one
orbit, be changed, and the whole stability vanishes.

Within the limits of the existing period we observe
a permanence of species, just as much as we find
evidence of it *within like limits* in ancient periods of
the earth's history; but beyond such limits of time,
and the influence of certain conditions accompanying
those changes, and differently affecting the several
species, we have no such evidence in the one case,
and direct *apparent* evidence to the contrary in the
other.

Argument
from con-
formity to
experience. But, secondly, the chief argument always is, that
we have " no experience of such a thing " as change

of species ; — that it is an arbitrary supposition " opposed to all experience ;" and the like.   Now the
argument of " *conformity to experience* " is beyond
question the very basis of all induction, but it
requires some caution in its application.   And on examining the argument, as here applied, we shall find
that there is a fallacy latent in the use of the
expression " we have no experience of such a
thing."   We may illustrate this by a familiar example.

We have " no experience " of the formation of
coal.   Yet in past epochs we know it occurred ; and
it is accounted for by known and existing causes.
The submergence of forests, — the accumulation of
vegetable matter, — the compression of materials by
superincumbent masses, whether solid or fluid, — are
known natural causes, which do, or might, occur
within our experience or that of history.   But for
the consolidation of those beds of vegetable matter,
and their conversion into coal, the essential condition
has been the influence of *immense duration* and vast
periods of past time ; and of this, undeniably, we
can have " no experience."

To apply this remark to the question of organic

Example from the formation of coal.

Immense lapse of

time essen-
tial to
changes of
species.

changes : It is alleged *we have no experience of such a*
thing as a change of species; but we have experience
of the present uniformity of species *subject to slight*
and occasional deviations. This is a known cause
now acting. To how great an extent these successive deviations might be carried in immense periods
of past time under changing external conditions, we
know not.

Thus this known cause, like that of the submergence of vegetable masses, *conjoined with the influence*
of incalculably vast periods of *past time*, MAY BE
fully competent to give results as remote from those
now *every day seen,* as the formation of coal has been
from what takes place in any submerged forest or
accumulation of vegetable matter in recent times.

Correct
statement
of the case.

Thus the *mutability of species* in past epochs
would not be a case impugning the doctrine of the
immutability of natural laws, or the appeal to experience of established physical causes in accounting
for past changes; because the very nature of the
case *essentially involves other elements* than any now
entering into our consideration. *General experience*
must be understood to include more than *present*
every-day experience; and the advocates of transmu-

tation do not assert it *absolutely,* but only in *indefinitely long periods of time.* They do not maintain that it occurs under existing conditions, but only under *great and peculiar changes of condition.* Any fair and correct statement of the case, then, must include these qualifications. *We cannot say that we have no experience of a change of species in a due length of time, and under adequate and appropriate changes of external condition.*

Thirdly, closely connected with the last argument, there is another consideration equally material. Supposing it true that we have *no experience* of a particular kind of event occurring, and supposing at the same time it could be shown that, if that event *did actually occur,* we could never (from the particular nature of the case) have any *evidence* of its occurrence, then it is clear the argument from *want of experience* must fall to the ground: it would be no proof whatever of the non-occurrence of the event in question. And, further, if there were a show of reason, from analogy, that such an event were *likely* to occur, still less could the absence of experience be urged as rendering it *incredible* or inadmissible.

Want of experience no argument where impossible to be had.

FF 3

Example:
planetary
systems
round the
fixed stars.

For example, we have " no experience," by observation, that any of the single fixed stars have planetary systems revolving round them.   But it is admitted that, if they had really such systems, no
possible telescopic power could ever show them to
us.   Hence the argument from want of observation
goes for nothing.   But farther, the stars are self-
luminous and analogous to suns; and by the same
analogy, they may *most probably* be surrounded with
planetary attendants.   On the whole, then, it is a
reasonable belief that they have such planetary systems, though all *experience* is, and for ever will be,
wanting to prove it.

Origin of
new species
could not
fall within
our ex-
perience.

Now, with respect to new species, by direct calculation, founded on a liberal assumption as to the
number of species, and the probable rate of their extinction, it has been shown* that such an event as
the extermination of one species, and the substitution
of a new one in its place, must be *an event of so rare
a character, that no noticeable instance of it could be
expected to take place within the range of our observation.*

This argument is independent of the supposed

---

* Lyell's Principles, p. 682., 8th edit.

mode of introduction of the new species. Hence it is to be observed, that it applies equally to the case of transmutation as to any other supposed mode of origination; and the consideration arising is clearly this— that if transmutation did really take place, *we could never expect to have any experimental evidence of it.* If it did occur, it would not fall under our notice : it is therefore no argument against it that we have " no experience " of it. The argument from want of evidence falls to the ground ; and the hypothesis *is not, on this ground, incredible or inadmissible.*

Argument applies equally to transmutation.

But, fourthly, still farther, let it be granted that such an event as the origination of a new species were really to occur at the present day, within the possible range of observation; let us imagine that some changes were actually going on within our own times so as in any way to give rise to a *really new species,* and we ask *how could the fact ever be substantiated ?* To say nothing of such facts occurring in the depths of forests and deserts, or in the recesses of the waters, where no human eye could by possibility ever detect them, let us suppose that something apparently of the kind were alleged to have

If new species did arise, no possibility of verifying it.

been witnessed by some competent naturalist, what could it amount to more than *the existence of some new peculiarity*, which, supposing it allowed to constitute the mark of a distinct species, could, after all, only be affirmed to be *newly discovered*.

Incredulity on such a subject.

Let us even suppose it were possible for such processes *to be watched and accurately examined*, how difficult would it be for even the most skilful and unprejudiced naturalist to feel quite sure of the real nature of the case! How much more readily would any other interpretation be put upon it than that of a really new, distinct, and permanent modification! or, rather, with what determined scepticism would not its reality be denied, and with what abuse and ridicule would not the unlucky observer be assailed, who should venture to assert the occurrence of such a thing as *an actual observation of the first beginning of a truly new species!*

Absence of intermediate links.

The transmutationists suppose changes in external physical condition, affecting the characters of species: these may be such as would tend to the extinction of a particular species; but some varieties of that species might possess peculiarities better suited to those changed conditions, and thus would be able to

survive. *These would be few;* but they would propa-
gate descendants in whom those characteristics would
be more strongly marked, sustained, and favoured by
the changed conditions. Thus for a period longer or
shorter, as the external changes advanced, the old
form would die out, and these rare varieties would
maintain a struggling existence, until at length, the
state of things becoming more settled, and the type
determined in accordance with them, a new fixed
species would begin to increase and multiply ; and it
would be of such common and wide-spread species
alone that we could ever expect to find fossil remains.

Thus, they would reply to the objection that *intermediate links and stages are missing—it is because
they were rare and transient.* The new species appear in company with other older forms not closely
allied, still persistent, because not affected by the
changes.

These considerations may suffice to show how very
little any reasoning from the mere *absence of evidence*
will really avail in the case before us. The objection
founded on it is, in fact, wholly groundless; and
there does not appear to exist any valid argument
to prove the general hypothesis of transmutation *in-*

Question
between
evolution of
organised
beings out
of their
inorganic
elements or
out of pre-
existing
forms.

admissible. Bu: viewing the question in a strictly philosophical manner, it necessarily brings us to the alternative, as a fair analogical conjecture,— if organic
life had a beginning, there must have been some stage
at which there took place *a first evolution of animal
forms out of inorganic elements;* and the question,
more precisely stated, then becomes, At various, repeated, subsequent intervals, corresponding to certain
epochs in the history of the globe, in order to give
rise to new species, did similar fresh evolutions take
place *out of inorganic matter?* or was it the case that,
when certain primitive stocks had been thus constituted at first, they were also subjected to certain
laws of modification of form, to come into operation
under the particular combinations of external conditions which were to mark future epochs, and that so
new species *were to be evolved out of the old?* The
choice between two such hypothetical ideas is a perfectly legitimate subject of conjectural discussion and
difference of opinion; but it is inconsistent with all
inductive principles not to admit that *one or the other
must be supposed.* But if the idea of a formation of
organised beings out of their inorganic elements
were to be preferred, still on any such hypothesis

the process is imagined to be carried on through such a series of steps of gradual evolution as to differ rather in name than in essential nature from the idea of development out of pre-existing organic forms.

But farther, even if the very cautious inquirer prefer *altogether to dismiss and ignore* the consideration of the question, on the alleged deficiency of satisfactory evidence, still, in the true inductive spirit, he admits that it is *nothing more than a mere physical question* which *at present* he cannot solve. And on the same grounds he would as strenuously contend against the admission of any hypotheses derived from other considerations, of a kind incompatible with the great principles of natural order, and of a nature beyond the domain of science.

If the question be ignored, it is still a physical question.

Every advance in physiological discovery seems to point to the necessity of an entire remodelling of the very ideas of *higher* and *lower* organisation, once so much dwelt upon. The structural relations of different species, more especially of what are called the *lower* orders, are disclosed with increasing indications of the real complexity of those relations; and we are thus, palæontologically, led to a perception of the higher connections which some of the

Law of succession of forms, complex.

earliest *observed* forms of zoophytic * life may claim ;
while the true idea of *progression,* in any sense, is
clearly no longer to be recognised in any single line
of ascent from more simple to more complex forms,
but must be sought in some new and apparently
less obvious train of relation not as yet made out.

All earliest   At the same time, all such speculations are de-
forms
destroyed.   prived beyond reparation of that first essential to their
completeness, a knowledge of what were *really the
earliest* forms of life on our globe, necessarily destroyed and burnt up as they must have been in the
metamorphic and igneous rocks.   The known fossil
flora and fauna of any formation constitute a mere
fragment of what, by all parity of reason, we must
suppose to have been the actual series of organised
beings of that period; and the aggregate of all these
known series are probably as small a proportion
of the whole of those which existed in still earlier
periods, and whose remains were in like manner

---

* As connected with these points, and more especially the develop -
ment of fishes, the reader will find some able illustrations, refuting
many of the minute details of the theory of the " Vestiges," in a very
able article in the British and Foreign Medico-Chirurgical Review, No.
xxvi., April, 1854, p. 425.
    But however open to criticism may be the details of particular phy-
siological statements in the work referred to, the whole tenor of the
preceding discussion will show the degree in which I cannot but concur
in its broad philosophical principles.

imbedded but, from the necessity of the case, have all been destroyed. And, according to the profound suggestions thrown out by one of our most, eminent philosophers, the destruction arose from the action of the very same causes which occasioned the elevation of strata, dependent on the effects of central heat.

Assuming that the earth is internally hot, and at a certain depth in a state of fusion, it follows necessarily that if in any part a great additional weight is laid on the exterior crust, the part below it will be pressed down, and, besides elevating the beds at the sides, will itself become more highly heated. This will be the case when, after a long series of ages, a vast deposit has accumulated at the bottom of an ocean : in proportion to the thickness of this deposit. will it press down the strata below it, till they become intensely heated, or even burnt and fused, and thus all the organic remains they may contain must be obliterated and destroyed. Every stratum thus deposited must have produced this effect on those existing before it, and as the author expresses it: "You see, therefore, that my object is to get at a geological '*primum mobile*' in the nature of a '*vera causa*,' and to trace its working in a distinct and intelligible manner.   In future, therefore, instead of saying, as heretofore, '*let heat* from below invade the newly-deposited strata (Heaven knows how or why), then they will melt, expand, &c., &c.,' we shall commence a step higher and say, '*let strata be deposited*, then, as a necessary consequence, and according to known, regular, and calculable laws, heat *will* gradually invade them from below and around, and, according to its due degree of intensity at any time, will expand, ignite, or melt them, as the case may be.'

"According to this view of the matter there is nothing casual in the formation of metamorphic rocks. All strata once buried deep enough (and due TIME allowed!!!) must assume that state, — none can escape. All records of former worlds must ultimately perish." *

Regular laws from unity of type.

But the invariable relation of all the successive forms to one primitive type constitutes the legitimate and undeniable evidence of some regular order of causes presiding over their production, operating

---

* Letter from Sir J. Herschel, in Mr. Babbage's 9th Bridgewater Treatise, Appendix, p. 240.   2nd edit.

through periods of time of enormous length, during which old species have slowly disappeared by the action of natural causes, and new allied species have as gradually appeared beyond all doubt as much in accordance with other equally natural, even if at present unknown, laws — parts of the great order of causes, in conformity with which these and all possible physical events must have taken place.

In what has preceded, it has been, I trust, sufficiently shown that some of the arguments most commonly adduced, whether in support of the " immutability of species," as supposed analogous to the permanent laws of nature, or against their " mutability " as " contrary to experience," or in favour of interruptions of natural order from apparent gaps in the geological series, are all destitute of foundation, fallacious, and untenable.

Recapitulation.

On the other hand, while these arguments, which are those most commonly relied on against transmutation, are in my opinion completely refuted, we must still remember it is but an hypothesis, there is still no *positive evidence* to establish it as *a demonstrated theory.* Yet as a mere *philosophical conjecture,* the idea of *transmutation of species under adequate changes of con-*

dition, and in incalculably long periods of time, seems
supported by fair analogy and probability.

No neces-
sity to sup-
pose inter-
ruptions of
order.

Taken for what it is worth as a conjectural hypothesis, it may be regarded as helping the general
conception of *some great principle of orderly evolution,*
according to which the present as well as past systems of existence have been produced out of preceding orders of things, and as at least conspiring
with all truly philosophical considerations to disprove
the necessity for appealing to any sudden interruptions of order, or operations of an unknown and
mysterious kind, alien from all natural causes.

Conjectural
hypotheses
in subordi-
nation to
broad prin-
ciples.

It should, moreover, be carefully observed that,
though particular hypotheses of this kind may fairly
be indulged in, regarded as such, yet an exclusive
devotion to any of them may be prejudicial to the firm
grasp which the mind should rather seek to maintain
of the *broad principle,* the subordination of all events
to *some general laws,* however at present *undiscovered,*
based on the maxim that, throughout nature, *what we
do not know must really be as much under the dominion
of law as what we do know.*

True science is always ready to confess the failure
of *existing* means of investigation when a limit

appears placed upon its advances, coupled, however, with an assurance that that limit will some day be passed.

It is legitimately within the province of inductive philosophy to suggest conjectures as to the operation of grander laws of vitality acting through the immense periods of past duration, and of which, during the brief and momentary duration of existing things, we enjoy only the most partial, imperfect, and occasional glimpses.

It is eminently consistent with the great principle of the uniformity of nature through all time, to suppose that like many lesser laws in the natural world, that of the *existing* permanency of species, may yet be subordinate to a greater and more comprehensive law of change, requiring such vast periods for its accomplishment that no measurable portion of time may suffice for the production of a sensible amount of variation.

Perma-nence of unity of composition.

In the *inorganic* world we recognise the order of scientific inquiry : first, chemical mineralogy examines the actual composition of the materials of the earth, and their distribution in the composition of rocks; next, geognosy points out the actual order of super-

Position of a theory of development in the order of sciences.

G G

position of these rocks; and lastly, geology, so far as merely mineralogical characters and mechanical arrangement are concerned, traces the action of that succession of mechanical causes which has given to the several beds their peculiar character and structure and has occasioned their relative order of superposition.

If we look to the *organic* world, the same ought to be the order and method of investigation. We have, in the first place, comparative anatomy and physiology determining the actual characters and species of fossil organic remains; in the second, palæontology classifying them according to the respective periods in which they existed, and tracing the epochs of their apparent origin, abundance, and decline.

Causal geology : Causal palæontology.

But the third stage of the investigation, and the science proper for it, is here as yet wanting : this is that branch parallel to that of *causal* geology in the inorganic department, whose province would be to investigate the physical causes which successively brought about those changes in species, just as depositions, subsidences, eruptions, upheavals, and the like brought about the changes in the inorganic phenomena.

To supply this deficient branch of science (as in

all parallel cases), some crude conjectural and tentative attempts would legitimately be, and in fact have
already been, made.   This is the place in the order
of sciences to assign to the theory of development;
which, as proposed by Lamarck, precisely fulfilled its
purpose of a first crude conjectural hypothesis.   In
its details, it has of course been subject to refutation,
and has been replaced by other like attempts having
the same strictly legitimate object in view; — but
endeavouring to remedy its defects ; while they may
themselves be still open to criticism on many other
points : they stand in the same relative position as
the various Neptunian and Plutonic theories did to
causal geology before the announcement of Lyell's
principle.   The science of *causal* palæontology remains
to be constructed, but the speculations alluded to,
however faulty *in detail*, are all just and philosophic as
first steps in the right direction : they are eminently
useful in indicating, in some degree, the course to be
followed, and in pointing more distinctly to the object
to be aimed at, viz., the explanation of changes in
species in ancient epochs, by the analogies of probable causes of such changes derived from actual
natural laws.

Having thus far considered the general principles, I will proceed briefly to glance at the more particular views of some writers.

Narrow
views pre-
valent.

Looking at the question in a perfectly dispassionate manner, there appears to me a one-sidedness in the censures, or at least, excessive cautions, often expressed, against so hazardous an hypothesis as that of transmutation, even by some eminent philosophers, more than is warranted by sober philosophical considerations; and in which others display more zeal than can be explained by mere antagonism in a fair scientific controversy, while they sometimes appear to betray a singular degree of alarm at the bare suspicion of a leaning towards the obnoxious theory of development, as if their whole scientific, or even personal, reputation were at stake.

Polemical
spirit of
some dis-
cussions.

Some, again, have taken up such questions in a more determined controversial spirit, and have maintained in a tone of polemical acrimony, little to have been expected on such a subject, that the phenomena of new species are absolutely impossible to be explained on any physical principles, or even by any physical conjectures; and must be ascribed to sudden interruptions of the order of nature, connected with the convulsions and catastrophes which overwhelmed all the old species, and were of a kind wholly beyond the domain of physical causes or the limits of philosophical examination.

Such imaginations easily find favour with those who have *some other object in view than mere philosophical truth ;* and if somewhat faulty in their foundation, their weakness in reason is abundantly compensated by loudness of dogmatism and a preremptory style of assertion that " species are real existences," and that "transmutation is impossible ; " all which has an imposing effect when supported by the aid of a kind of mystified eloquence, and seconding the more awful denunciations so authoritatively pronounced against the heterodox speculations of the developmental school.   <small>Dogmatic assertions.</small>

But when (as we have observed) some of the opponents of transmutation do not content themselves with mere negation, but assert another theory of sudden originations of animal life, we have clearly a right to demand of them some distinct statement *of their own meaning,* some definition of the nature of the theory they propose to substitute, and the *process* by which *they* conceive the results to have been brought about.   <small>Theories of sudden origination.</small>

Let us, then, imagine the case in question ; among other allied species already existing let us suppose that a really different *and truly new species* has *suddenly* made its appearance.   The individuals of this new species are found living and growing by ordinary means, and in all respects subject to the same regular conditions as everything around them.   It is the fair right and object of the inductionist to ask *how long* have they continued in this state ?   Did the natural process of growth reach back to their evolution from a seed or an ovum ?   or to *what stages* of early existence or rudimentary evolution ? *

Altogether fanciful and unintelligible.

If not derived from a parent, was the ovum formed out of its component elements already existing in matter around ?   Was the organism gradually evolved, or do those who adopt a different view really mean

---

* Mr. W. J. Hamilton (Address, Geological Society, 1856) questions my right to demand of my supposed opponents, a statement of how they imagine species to have originated : but it will be seen on referring to the text that all I demand of them is a " statement of *their own meaning* " — in asserting a sudden origination which appears, on the face of it, to have some meaning quite apart from any properly scientific theory.   He also retorts the same demand on me, but that demand *is already answered.*   I have already stated my view of the matter, which may be described to consist in regarding the evolution of new living forms to have taken place *undoubtedly* according to *some determinate physical laws* — if as yet unknown : — *possibly* by some process of transmutation from existing organised existence, as a not unphilosophical *conjecture :* or perhaps by origination out of *inorganic* elements : but at all events, viewing the whole question if ignored as incapable of *present* solution, yet properly left as a *physical question* for future inquiry. *Vide* p. 423.

that it assumed its form suddenly, or that the entire creature started into existence, full grown out of the earth, as in the frescoes of the Vatican, or the imagery of Milton*; or that it assumed a palpable existence out of nonentity ?

If such be their meaning, all that the inductive philosopher asks is to see some slight *proof* for these marvellous assertions, *which he will be quite prepared to admit if sufficiently verified;* but must, after all, ascribe to some action of regular physical causes as yet unknown. The " onus probandi " clearly lies on those who assert such extraordinary hypotheses, and not on those who, if they indulge in any speculations of the kind, are careful to found them strictly on probable analogies, dependent on the great laws of unity of composition and modification of parts; but in every case strictly conformed to the one grand overruling principle, the universality of law, order, and continuity, presiding as powerfully over the earliest stages of creation as during its continuance at the present moment, and applying equally to organic as to inorganic existence.

Necessity for proof.

---

* Par. Lost, vii. 463.

Inconsistency of such views.

Thus some scientific inquirers reject the idea of development or transmutation, because, as they allege, they find no evidence or existing instances of such a thing to produce; yet, in its place they assume a sudden production of full-formed animals out of their elements, or out of nothing, which is still farther remote from any possibility of proof from experience, and even beyond all rational conception. They discard one theory for want of proof though easily imagined and understood, only to adopt another which equally wants proof, and is at the same time wholly unnatural and incapable of comprehension.

Fallacious illustrations.

It is but putting the fallacies above refuted into a variety of vividly illustrated forms which constitutes the staple of some very popular writers. We find it, for example, pervading the work of Mr. H. Miller * (already referred to). To take a single instance, we may cite that striking passage where he so graphically describes the lake of Stennis, partly salt, partly fresh, partly brackish, and observes that each portion has its appropriate species of plants and animals, which

---

* Footprints, &c., p. 240.

never have been found to exhibit any signs of change, migration, or intermediate mutations, *so long* as they have *been observed,* and the *conditions* of its several parts have remained the same; whence we are to conclude that no such mutations ever are, have been, or can be made, under any conceivable changes of condition, and during any period of time, however incalculably great!

The same writer upholds the truth of the gradual " elevation of the types of being in the successive stages or conditions of the earth, as corresponding to those changes in the dwelling-place assigned to the animal creation, and suited to their successively improved natures."* And it is his main object to maintain that these changes were all of an isolated nature, and that they all happened in an abrupt and unconnected manner, inexplicable by natural causes; and this (it would seem) solely on the ground of the alleged breaks or interruptions (as he considers them) between different formations, which were before considered.†

As to the mode in which these changes were

---

* Footprints, &c., pp. 283. 286. † See above, p. 356.

brought about, the author appears to proceed on the
ground that *all* scientific modes of explanation *must
fail,* because *one* such mode (in his opinion) fails; he
therefore takes refuge in the assertion of immediate
and repeated sudden *" creations,"* as the only way in
which the continued production of new species ought
to be spoken of, and even strongly denounces all
attempts otherwise to explain them.   And (unless I
misapprehend the author's meaning) it would seem
as if he seriously upheld the notion of animals being
thus produced *full grown,* since he expressly considers Oken's theory of development out of a monad
or infusorial point, sufficiently refuted by referring to
the existing *size* of the gigantic Asterolepis *, and
other fossil remains, which he assumes to have been
thus suddenly produced; as if, because they had *grown*
to that size, these creatures had never been in an
embryonic state, perhaps microscopically minute.

Physical
order in
all past
changes.

Legitimate science can never lead us to anything
but higher generalisations of physical order; it can
never point to operations of a kind beyond regular
causes, or warrant a reference to hypotheses stamped

* Footprints, &c., p. 119.

with the professed character of mystery and inscru-
tability, at an earlier, any more than at a later epoch;
in the primeval arrangements any more than in the
existing maintenance of the organised world.   Where
we confess ignorance of intelligible truth, it is at once
absurd and presumptuous to create inscrutable mysteries, and to put them forth as the conclusions of
science.   It is clearly preposterous to maintain that an
ancient event ill understood is an absolute deviation
from all natural order, merely because we cannot at
once interpret it; or that it is beyond all physical
causes, because we do not daily see instances of its
occurrence, by the action of such causes.

But, lastly, if it were granted that we could follow
up the successive development of species even to a
very remote date, it must be evident that this would
still carry us but a little way towards *the real first
origin of all things,* and would manifestly be but a
single step in the course of tracing backwards the
order of creation.   The question of the *first origin
of organised life* would still remain.   In reference
to *successive forms and changes* of life we have
the evidence of existing remains as the basis of our
reasonings.   But we have no such evidence of the

Explana-
tion of
changes no
explanation
of first
origin.

beginning of the series. Even if the idea of transmutation or any equivalent principle were granted
as to comparatively later variations, still the question
would arise, What and how many were the original
types from which these unlimited varieties began to
diverge? And how did the primitive germs of life
themselves originate? Yet even on these points
theories have not been wanting.

Theory of
Lamarck.

Lamarck represents nature as continually engaged
in the gradual formation of the elementary rudiments
of all animal and vegetable existences of the simplest
kinds, which are afterwards compounded into more
complex forms. These rudiments or monads are the
only things to which she gives birth directly. He
regards them as probably of a distinct kind for each
of the great divisions of the animal and vegetable
kingdoms, and supposes that they are gradually developed, but subject to material modifications from
the action of external causes.

Theory of
Oken.

This theory is closely allied to that of Oken*,

---

* That remarkable work, the "Elements of Physio-Philosophy" of
Oken, has attracted some notice in this country, through the translation
of it published under the auspices of the Ray Society (1847). Its
nature is confessedly speculative and hypothetical. The author expressly
tells us in the preface that he wrote it off under " a kind of inspiration,"
and more than once refers to its deficiencies in proof from matter of fact.
With regard to the primary principle of organised life the author de-

which, avowedly of a speculative kind, and bearing a
metaphysical aspect in its first principles, is yet put
forth as a physical generalisation ; though from the
very abstract nature of the ideas and language employed, it is difficult to estimate the precise evidence on
which it is supported.   The author's main principle,
however, appears to be the origin of all organised
life from a primary infusorial cell or monad, formed
out of an elemental substance, which seems to be
simply a compound of the admitted inorganic elements
of animal matter.   From the aggregation of such
cells the various organic structures are compounded,
and this he seems to regard as a process constantly
going on in nature.

It is sufficient here to glance at such speculations,

---

clares it the object of his speculations to show how, by self-evolution of
the elements into higher and manifold forms, they become finally organic,
and in man attain to self-consciousness.   Man includes the representation of all lower forms, and these again are but man disintegrated.
(§§ 10—19.)   The grand principle is the origin of all organised life from
an infusorial cell, formed out of what the author terms the " primary
mucus," " schleim-substanz," or protoplastic matter, which, he says, is
" carbon mixed identically with water and air." (§ 898.)   Decomposition is only a transition from one life to another, which takes place
through this mucus, into which organised matter is redissolved. " Every
generation is a new creation." (§ 924.)   Of this mucus a cell, cyst,, or
vesicle, is formed, called "infusorium," the primary germinal principle
in plants and animals. (§§ 930—943.)   The author says, " No organism
has been created of larger size than the infusorial point, which is microscopic: all larger forms are developed, not created.   By the aggregation
of such vesicles organised forms arise, which by successive combinations
of the more simple, produce ultimately the higher and more complex
structures." (§§ 958. 2961. 3161. 3175.)

all essentially hypothetical; these and others of the like class are at least of use if they only serve to show that there is nothing *impossible* or *inconceivable* in the idea that such production of the rudiments of new life may have gone on in accordance with a regular physical system.

Inductive inquiry does not extend to a beginning of all physical causes.

However high we may ascend in the order of time, it is still perfectly within the province of physical inquiry to endeavour to unfold the steps in the process by which changes in the order of things have been brought about. It is within the limits of philosophical conjecture to speculate on the gradual accomplishment of the great design of educing the existing order of the universe out of former conditions, in whatever degree of obscurity the precise *modes of action* may be enveloped by which it was effected.

As far as we can trace backwards, even in imagination, the succession of events into the depths of primeval time, we can only conceive them succeeding one another in determinate order, by the operation of profoundly adjusted causes, whose nature becomes less and less imaginable to us as we recede into higher antiquity; but we cannot draw any line, and say here the series began, or here all continuity commenced.

To attempt to go back in imagination to an epoch prior to all the changes of matter at which it was first constituted or called into being, is what no inductive philosophy can warrant. No analogy points to such a beginning of physical causes. Physical philosophy always supposes at least some physical elements in existence ; it cannot investigate or conceive a condition antecedent to nature, or the case of its actual commencement. No science can carry us, even in imagination, into a state of arbitrary and disordered influences ; a *chaos* has no existence in the ideas or the vocabulary of the inductive philosophy. A creation, in the same vocabulary, implies orderly evolution. If we entertain any ideas beyond these, it can only be from sources of quite another kind.

All ideas of a beginning from other sources.

§ IV. — THE BEARING OF THE PRECEDING ARGUMENTS ON THE THEOLOGICAL VIEWS OF CREATION.

Theological view of the inquiry kept distinct.
IT was observed at the outset of this Essay, that the question of creation has distinctly a theological bearing ; and it is no disparagement to such a view, that in the preceding sections I have treated the subject in a purely inductive and scientific light, and have *purposely abstained* from introducing any refer‑ ence to those higher considerations, in order to lay a more secure basis for any such applications, as well as for meeting any objections alleged on religious grounds.

Prejudice against such specu- lations on religious grounds.
It cannot be denied that any discussion of the question of Creation, or any attempt to trace the probable history of the origin of the physical world, or of its organised productions, on merely scientific grounds, has been often regarded, especially by a certain class of minds, as having a tendency unfa- vourable to religion, and as being, in some degree,

an intrusion into its province and an assumption of
its office.   Such impressions, however, appear to   Arising
                                                      from mis-
me to take their rise in the same common species of   conception.
misconception of the relations in general between
science and faith, which, in so many other instances,
has resulted either in a lamentable antagonism and
hostility, or in futile attempts to combine them in
incongruous union, upon fallacious principles.

I have, in another place *, considered some instances
in which the discoveries of science are undeniably at
variance with doctrines which had become identified
with popular belief, or had even been erroneously
received as part of the established creed.   And
when any topics having a similar bearing, come
into discussion, if, on the one side, they are naturally
taken up with the zeal of religious prepossession, on
the other, they are often not fully examined ; they
are impatiently dismissed, or thought sufficiently
treated, if glossed over by a few vague, specious, and
evasive generalities.

There exists, unhappily, too great an unwillingness   Want of
                                                      open dis-
on either side to meet such questions with perfect   cussion.

---

* See Essay II. § ii.

H H

honesty and fairness. The astronomer, the physiologist, or the geologist, for example, may be fully
enlightened as to the extent to which some of the
conclusions of his own science may clash with certain
received articles of popular belief. But, devoted to
that science, and caring more to relieve it and even
himself personally, from hostile insinuations, than to
promote any higher views of truth, he more naturally
than philosophically seeks to conciliate the matter
in an ambiguous phraseology; as if accepting literally
the irony of Lucian, who, after relating a story of a
philosopher having been maltreated by a mob for
attacking some of their superstitions, adds, — " And
very justly; for what right had he to be rational
among so many madmen? "

Suppression of discussion injurious to religion.

But still more injurious to the cause of religious
truth is the course too often resorted to by the professed defenders of its cause, even in the present
times. Not always duly alive to the actual spread of
intelligence, they cringe to the loud but ignorant
zeal of the few, and become followers in the train of
prejudice rather than its correctors and enlighteners.
They have too often yet to learn that, by continuing
to insist on dogmas which the advance of knowledge

has discredited, and literal interpretations which the discoveries of science have set aside, by adopting fallacious compromises, or by discouraging and denouncing those open avowals which alone consist with the reality of truth, and that free inquiry which Christianity challenges— they are following a course as unworthy in principle as it is short-sighted in policy; they are inflicting the worst injury on their own cause, and are but strengthening the arms of that sceptical hostility which they so strenuously profess to oppose.

On the other hand, here, as in other subjects, resolutely yet cautiously, to pursue the free course of rational inquiry — and therein to follow *truth*, — we may be assured can never lead to *evil;* while every advance in real enlightenment, in proportion *as it is real*, must of necessity cast its beneficial rays equally over science and over faith.

In a former Essay* I have, it is to be hoped, sufficiently shown how groundless is the ignorant, but common, prejudice that a reference to orderly evolution and physical changes occurring according to de-

Evidence of a moral cause from physical causes.

---

* See Essay I. § v.

terminate laws is at variance with the belief in
design,—of which these laws are the very indications; that the appeal to physical causes has any
tendency to exclude a moral cause, — of which they
are in fact the evidence; or that the higher we trace
the series of such causes, the further we postpone
the recognition of a Deity,—of whom that series of
causes is the very manifestation. And if we look back
to the order and succession of physical causes, and
trace the process of physical evolution in past epochs,
we obtain continually enlarging and accumulating
proofs of the same sublime inferences.

Evidence of
science to
the su-
preme
mind, not
to the act
of creation.

In the preceding discussion, we have seen that
physical inquiry traces the development of the existing material world up to a certain point, and allows
us to conjecture a few stages beyond that; but
where analogy ceases to apply, or conjecture to find
materials, there all *physical speculation simply terminates :* it does *not* even point, however obscurely, to
any event beyond. Its province is *Nature* in space
and in time. With any class of ideas *of a different
kind* not referring to *Nature*, it has no concern. The
inference, indeed, from all physical truth which impresses us with the idea of *Omnipresent Mind*, is one

everywhere presenting itself, as much in every phenomenon at the present moment, and in our actual locality, as in the remotest points of distance in time or space.   The physical evidence of *Creation* is universal, but there is no indication of any one point as its commencement.

The proofs to which I have referred in a previous Essay * of the great Moral Cause of the universe, are of a kind related rather to *permanent* and enduring evidence of the order of things, than to any inferences as to special *past acts* or *events*, which may be imagined to have been its direct manifestations at remote epochs ; but any such supposed events would afford similar evidence only in proportion as they might be found evincing a conformity to some high principle of order and law.   The evidence of the material world points to *ever-existing Mind* continually manifested in the existing order of nature, as well as in all the successive changes which we can trace through countless periods of past time ; but which, in all their varied modifications, present not the smallest deviation from one great type of unity and harmony.

---

* Essay I. § v.

The question of " creation " in any philosophical
sense, is closely connected with the view we take of
" causation." According to the view here adopted a
physical philosophy traces " causes " in more *generalised relations of antecedence,* not in any idea of
efficient power.* It teaches us nothing of causation in
the sense of active agency. It traces causes of existing phenomena, " laws of laws," but not in the
sense of originating power; such an inference is not
within the bounds or scope of inductive science.
From physical philosophy we neither have nor can
have any evidence of a beginning of physical causes.
The true argument does not rise to the idea of origination of laws or production of being. The real and
undeniable inference is that of universal reason or
intelligence pervading all nature; and this not so
much a conclusion as the only language in which
it is possible to express the facts.

It is this inference extending through all past duration as well as the present, which constitutes the
true evidence of natural theology, and not any idea of
commencement or origin.

---

* See before, Essay I. § v.

Yet this view has appeared to some injurious to natural theology, but only because they cannot divest themselves of a lurking and unconscious adherence to the notion of efficient causation as connected with its truth.

Solely from the same erroneous notion, is derived the whole force of the objection before alluded to, which, though often repeated in various forms, is identical with that of Hume*, that from the indications of design we cannot infer the creation of the world from a *personal agent*, as we do in the instance of human works from the like indications.   This, however, in no way affects the real argument for mind and reason in nature.

Personal agency.

The same argument has been sometimes differently expressed by saying that we cannot infer an act of volition as in like cases originating in human volition : to which the same remark will apply.†

Yet numerous have been the attempts (as those of Chalmers and others) to reply by arguing for such origination on the old and unphilosophical notion of

---

* Dial. on Natural Religious Works, ii. 446.
† See Mill's Logic, i. 371.

efficient causation. On the grounds here advanced, these attempts are as needless as they are unsatisfactory.

M. Comte objects to natural theology that it makes all phenomena dependent on volition, thus *essentially arbitrary* and *irregular :* as indeed he might suppose from the mode of statement commonly adopted.

Whereas the view here taken shows, on the contrary, that the real inference of natural theology essentially depends on the fact that all phenomena are *invariably* and *universally constant* and *regular.*

Belief in a beginning from other authority.

But what is here remarked in no way disparages such ideas, whether of origination, personal agency, or volition; but merely shows that to whatever extent they are entertained, they are not ideas of philosophy, but derived from other sources. In a word, the idea of *a beginning of Nature in time* is one which no *physical* philosophy can teach us. It is an idea wholly deduced from other considerations.

Metaphysical argument.

There are metaphysical arguments as to the impossibility of conceiving eternal matter, because (it is alleged) it must then be self-existent, and the like. Such arguments turn on the impossibility of conceiving matter either to possess in itself a principle of eternal existence, or to be the originator of its own existence. But to avoid this difficulty by inferring its origination from another self-existent being, is only to involve ourselves in greater metaphysical difficulties; it being equally beyond the power of the human faculties to conceive *any* origination of matter out of nothing : still more to conceive self-existence at all. All arguments of this kind, however imposing in appearance, when strictly analysed, are found to involve ideas really beyond the province of human reason. In point of fact, by far the mass of mankind have obtained their idea of " Creation," not from any such arguments, but from the prepossessions of early instruction, by which that term, with a certain religious meaning affixed to it, has from childhood been incessantly impressed on their ears and memory,— though often but little deeper.

Thus, that from any testimonies of science we obtain but a very unsatisfactory *philosophy* of creation must be fully admitted ; in fact, we have, strictly speaking, no philosophy of *creation* at all ; we have only a philosophy of a vast and illimitable series of *changes*. We cannot extend our view to an actual first origin of things.

Philosophy and theology of creation.

The case is quite different when we turn to assertions which convey *not* the philosophy, but the *theology*

of creation.   But these are entirely distinct, and rest
on a totally different kind of authority.   In a theological view, we cannot obtain any other ideas of the
subject than are conveyed in those modes of expression, neither borrowed from inductive philosophy, nor
leading to anything in philosophy, but derived wholly
from other and higher sources, and to be understood
according to the design intimated.*

Nebular
theory ac-
cordant
with the
idea of
design.

As to any speculations of science on the earliest
history of the formation of the world, such as the
nebular hypothesis, or any like supposition, in
reference to our present view of the subject, there is
only this to be said: in proportion as any such
theory (supposing for the moment it were established)
suggests an *orderly evolution* according to *preordained*
physical laws, *so far* it keeps in harmony with the
idea of *Supreme Intelligence,* and thus implies a
worthy notion (as far as a mere conjecture may
avail) of creative power and wisdom.   If we could
in imagination trace the supposed process through
any antecedent stages in a similar way, it would in

---

* These remarks will, I trust, sufficiently remove some objections to
my views, raised in a very able article in the Westminster Review, July
1855, p. 217.

proportion afford a similar extension of such reflections.

It suffices to allude to such hypotheses in connexion with the present subject, merely for the purpose of remarking, on the grounds now referred to, how entirely futile and irrational are those charges often brought against theories of this kind, that they have an irreligious tendency.

When from such remote contemplations we descend to the comparatively more accessible inquiry into the *changes* which have taken place in our globe and its organised inhabitants, it is fully admitted that the farther these successive changes have been investigated, and traced to determinate laws, — the more links in the chain we can unravel, — the more they are always found to disclose the evidences of creative wisdom and design.   But, when we speak more precisely of " Creation," in the sense of the commencement of organic life — and especially of the introduction of new forms of life, — it becomes more necessary to examine the bearing of the theories of their origin on the cause and argument of religion, whether natural or revealed.

Since, on the high principles just referred to, the

Subsequent general arrangements of our globe indicative of design.

Views of origin of life and of new species in relation to the same argument.

No real interruptions.

truly inductive philosopher cannot suppose any such idea as that of a real deviation from unity of plan, so he cannot but feel assured that such an event as the introduction of a new species, or even the first origination of life, could be nothing of an arbitrary kind, but must have been part of the great order of preordained causes whose pervading influence essentially distinguishes CREATION from CHAOS; and in which any apparent interruptions can only arise from our confined apprehensions of the vast scheme, essentially *one.*

Unity of plan through incalculable periods of past time.

The evidence of palæontology throughout such inconceivably vast periods, is of the most overpowering force. The simplest contemplation of the facts of ancient organic life shows that even those forms of the earlier epochs which are the most dissimilar to any existing are all of *one family* with them. Throughout these unfathomable depths of primeval time which it transcends imagination to conceive or arithmetic to express, the organic world is, and always has been, emphatically *one :* modelled on one plan, and amid all diversity exhibiting one common feature of a grand *recondite and comprehensive unity of design ;* or, as Œrsted has expressed it, " The animals and plants of former periods are all different emanations from the same great thought."

Equally emphatic are the opinions expressed on this point by some of our most eminent men of science.

In continuation of a forcible passage before quoted, Professor Owen goes on to observe, "To what natural laws and secondary causes the orderly succession and progression of such organic phenomena may have been committed, we as yet are ignorant. But if without derogation of the Divine Power we may conceive the existence of such ministers, and personify them by the term *nature*, we learn from the past history of our globe that she has advanced with slow and stately steps, guided by the archetypal light, amidst the wreck of worlds, from the first embodiment of the vertebrate idea under its old Ichthyic vestment, until it became arrayed in the glorious garb of the human form." *

To this noble passage I cannot forbear adding the single comment that, according to my view, not only " without derogation of the Divine Power," may

---

* On Limbs, p. 86.

we entertain the ideas so beautifully expressed; but, if there be any truth in what has been before advanced, so far from anything *derogatory*, such a view constitutes the *very proof* and manifestation of that power, and is just what enables us legitimately to trace its operations—as alone we can worthily trace them—in the indications of law and unity, order and system; while without such evidences of Universal Mind and Supreme Reason, arbitrary intervention might be only irresistible fate, and sudden revolutionary changes and convulsions only atheistic anarchy.

Again, it would, perhaps, be impossible to find a more truly admirable exposition of the case than in the following sentence from the pen of one of our first living philosophers:—"For my own part I think it an inadequate conception of the Creator to assume it as granted that His combinations are exhausted upon any one of the theatres of their former exercise; though in this, as in all His other works, we are led, by all analogy, to suppose that He operates through a series of *intermediate causes;* and that in consequence, *the origination of fresh species,* could it ever come under our cognizance, would be found to be a *natural*, in contradistinction to a *miraculous*, process; although we perceive no indications of any process actually in progress which is likely to issue in such a result." *

As in the natural world the only indications we have of the operations of the Divine mind are those manifestations of order; so whatever we ascribe to the same source we can only conceive as worked out in accordance with the same great principles.

Evidence of creation in physical evolution.

The real question in any such cases is not whether certain events or processes are or are not to be traced to the Divine will and counsels — for that is not denied by any reflecting inquirer, — but simply whether the *mode and method* of the Divine operation can be either absolutely discovered, or even reasonably conjectured, to have proceeded in this or that particular path.

Question not as to the fact, but the mode of Divine operation.

A rational physico-theology teaches that the succession of forms of organised life on the globe, up to the first origination of all animated nature, were acts of the Divine will, wisdom, and power, in precisely the same sense as the revolutions of the double stars

---

* Letter from Sir. J. Herschel, in Babbage's 9th Bridgewater Treatise, Appendix, p. 226. 2nd edit.

and planets, the daily tide, the fall of rain, the ascent
of vapour, the action of the sun's light and heat, and
all other natural phenomena, regulated by similar
recondite laws, are direct and immediate acts of
the same Divine will, wisdom, and power.

Creative
power con-
stantly
manifested.
And, indeed, to approach still nearer to the idea
of *origination* or *production*, we may find creative
power as strictly and properly exemplified every
day in the marvellous process of evolution of animals
and plants out of a mere microscopic germ or
embryo, as in any events of past times. Those
events may perhaps appear preternaturally magnified
to our intellectual vision from the medium of unfathomable antiquity through which we view them;
but we may be assured that simplicity is as sure a
mark of the Divine operations as grandeur; and
that equally in the present as in the remotest epochs
of the past.

Professor Sedgwick (in a passage partially quoted
before, p. 415.) observes: "If it be affirmed that
the origin of the organic world was determined by
law, we believe the proposition true; partly on
the strength of what seems sound analogy; for if
the organic world be governed by law, we cannot

believe that it commenced without law : partly on
its obvious adaptation to the existing laws of the
organic world ; partly also on the ascertained historical
development of the forms and functions of organic
life during successive epochs, which seemsto mark a
gradual evolution of creative power manifested by
a gradual ascent towards a higher type of being."

Understanding the term " creative power " in the
sense before defined, as simply expressive of our
ignorance on scientific grounds of the *mode of origina-*
tion of organic life and its varieties, and allowing for
a doubt as to the *progressive* scale, I thus far entirely
agree with the able and eloquent Professor.

He continues, however: " But when it is affirmed
that the successive parts of the great organic sequence are related to one another only in the way
of material cause and material effect, we test the
proposition by an appeal to facts and experiments —
the last appeal on all questions of natural science —
and on the strength of this appeal we deny the truth
of the asserted proposition. "

But, as above shown,* in this instance *we cannot*

---

* § iii. p. 434., *et seq.*

make the appeal to experience: and because we cannot reproduce the conditions: thus the *absence* of
evidence *proves nothing* either way: the sole appeal
we can have is to general analogy, and the broad,
undoubted, incontestible ground of the continuity
and uniformity of nature.

Again, in a subsequent sentence, the author observes: " Those who exclude from their *creed* all
conception of a personal and intelligent God in
nature, must believe that dead inanimate matter
may, without external aid and by its own inherent
powers, work itself into what is vital, sensitive, and
intellectual." In this, again, I entirely concur, provided we keep strictly to the distinction that the
belief referred to is part of our " *creed* "—that is, is a
matter of " faith," not of *science*. But considered as
a point of physical inquiry, there is nothing contrary
to legitimate analogy in supposing vitality infused
into dead matter under certain preordained combinations of conditions: still less in imagining that
vitality continued under changed forms of organised
matter.

But there are some particular instances which have
been often dwelt upon as putting the question to a
more distinct test, and which may be perhaps more

peculiarly suggestive to those who have not been prepared to take the more enlarged view of the subject on first principles.

To take only a single instance, rather by way of illustration, we may refer to the multitudes of infusoriæ and animalculæ whose existence is *restricted and peculiar to artificial products* made by man, and must have been consequently introduced at dates not only *subsequent* to that of man, but continually recurring.

Now, the question of their origin involves equally remarkable consequences whichever way it is viewed. If they are " developments " of existing allied species, only modified so as to suit the particular conditions under which they exist, a principle is conceded which cannot be consistently refused in other cases. If they are the result of special interventions to bring them into existence out of nothing, they constitute such a multiplication of miracles as the most strenuous advocate must disavow ; and after all, according to all acknowledged principles, a miracle *continually* and *regularly* repeated ceases to be a miracle.

Theories of the physical evolution, or origination,

Instance.

Objections

to physical
theories of
evolution.
of new species in past epochs of the earth's history have been specially the objects of censure and denunciation to some eminently religious writers; and they have been sometimes condemned with a degree of warmth and violence which clearly indicates the admixture of a larger element of religious prepossession than of reason.* But even in a religious point of view it is, in truth, by no means easy to see on what substantial grounds such vehement opposition can have arisen.

Objections
unreason-
able.
So far as the evidences of *natural theology* are concerned, it would follow, from all that has been here advanced, that those evidences, so far from being discredited, could but receive increasing confirmation in proportion as *any physical theory* might be substantiated, which would give us a deeper insight into those secondary processes and laws by which the Divine

---

* Such accusations are often very unfounded. But when they are imagined subservient to a sacred cause, truth and fact, justice and candour, are too commonly looked upon as secondary considerations, or rather the disregard of them would seem to be considered as only the more praiseworthy evidence of religious zeal.

One of the most striking instances of this kind of religious vituperation at the present day has been that of the incessant attacks made on the "Vestiges of Creation," on the alleged ground of the impious and atheistic tendency of its speculations.

Now whatever may be thought of the theory or speculations of that work as such, nothing can be more utterly and palpably unjustifiable than the charge of an *irreligious* tendency against a work *in which almost every page is replete with expressions of the most devout homage to the Divine power, wisdom, and goodness.*

will and plans were worked out, and which are to us the manifestations of such designs.

But, perhaps (as in other instances elsewhere noticed*), nowhere has the confusion of thought respecting " causation " been more misleading than in reference to the subject of *Creation*, giving rise to the constantly reiterated but absurd accusation against all theories which aim at tracing the series of physical events as far back as possible towards the origin of things, that they fly to " second causes " in the desire to avoid the acknowledgment of a " First Cause," or that they endeavour to get rid of a Creator because they seek to trace more in detail the steps by which His work is carried on. Confusion of ideas.

Such often refuted cavils are, however, constantly revived by a certain class of minds; indeed, every advance in discovery or philosophic speculation from the days of Galileo downwards, has been, as a matter of course, accused of having an irreligious tendency.† Bigotry.

---

* See Essay I. § v.

† With a certain class of religionists every invention and discovery is considered impious and unscriptural—as long as it is new. Not only the discoveries of astronomy and geology, but steam, gas, electricity, phrenology, mesmerism, political economy, have all in their turn been denounced; and not least, chloroform. Its use in parturition, with admirable sense and consistency, has been anathematised as an infraction of the penalty pronounced on Eve!

Arago mentions that when the equation of time was first introduced,

Every successive step made good in bringing Nature under the dominion of law, is stigmatised as setting up Nature instead· of God, — as if we could trace Him except through Nature; and as referring everything to mechanical causes instead of the Divine will,— as if the recondite system of physical causes were not the very evidence of that Will combined with Supreme Wisdom.

The same objections must apply to all physical theories.

And we may observe, that if the supposition of original adjustment superseding continual interposition be objectional in the instance of a succession of varied forms of species, evolved by some processes and in accordance with some fixed law as yet unknown, it must be equally so when evinced in other instances better understood; for example, in the " stability " of the planetary system, the " conservation of areas," or even the very "inertia " which keeps up the planetary revolutions, or, indeed, throughout the whole system of the physical world referred to an invariable system of laws originally imposed, and by the combinations of which all actual phenomena are brought about. If perpetual inter-

and the clock affirmed to be more true than the sun, it was denounced profane.

vention and constant new volition be the only religious view of the matter, we ought to discard Laplace and Newton and go back to Kepler's "vital forces" and spiritual beings spinning the planets in their orbits, or Descartes' World of Vortices animated by the Divine Soul.

Others have denounced all theories of physical evolution as leading to and implying *Pantheism*, an accusation, if possible, more strange and groundless. Even in a theological sense, the question between development and successive "creations" is simply whether the Creator be supposed to construct a machine which, once adjusted, shall go on fulfilling its work, or one which at successive periods shall require repeated manual interposition. But the assertion that the universal machine is so constructed as to require no interposition, has really nothing in common with the Pantheistic theory, which (to carry on the same metaphor) would assert that the machine is not only self-animated, but is itself the artificer and source of its own parts and movements.

Objection as leading to Pantheism.

But the main source of the difficulty and objections which have been felt, on religious grounds, against any theory of the evolution of organised

Objections to development theory on doctrinal and

scriptural grounds.

existence by the agency of natural causes, is their supposed repugnance to particular views of theological *doctrines*, and the declarations of *Scripture* on which those doctrines are founded.

Discrepancy between Scripture and geology.

In a former Essay * I have adverted to the question of discrepancies between science and the language of Scripture generally, and have referred more especially to that notable instance of it — the irreconcileable contradiction between the whole view opened to us by geology, and the narrative of the Creation in the Hebrew Scriptures, whether as briefly delivered

Even where admitted, other doubts remain.

from Sinai, or as expanded in Genesis. In the minds of *all competently informed persons* at the present day, after a long struggle for existence, the literal belief in the Judaical cosmogony, it may now be said, has died a natural death. Yet many are still haunted by its *phantom*, which perpetually disturbs their minds with apprehensions equally groundless, on collateral points.

Most rational persons now acknowledge the failure of the various attempts to reconcile the difficulty by any kind of verbal interpretation; they have learnt

* Essay II. § ii.

to see that the "six days of thousands of years" have, after all, no more correspondence with any thing in geology than with any sane interpretation of the text. And that the "immense period at the beginning," followed by a recent literal great catastrophe and final reconstruction in a week, is, if possible, more strangely at variance with science, Scripture, and common sense. Yet, while they thus view the labours of the Bible geologists as fruitless attempts, they often do not see that they are fruitless, not because they fail in detail, but because they proceed altogether on wrong grounds and in a wrong direction, and thus remain under the dominion of the same radically mistaken prepossessions, which lead to not less unhappy misconceptions on other allied topics.

Well might Humboldt* speak of geology as "now finally abstracted, *on the Continent, at least*, from Semitic influences." But in this country it may be hoped a better epoch is beginning to dawn, as it must do, in proportion as men reflect on the real basis of their reasonings, and learn to apprehend

---

* Cosmos, 1st transl. p. 288.

clearly the distinct grounds on which science and Christian belief respectively repose.

Evolution not more opposed to Scripture than all geology.

And if to the general truth of the immense continuous series of slow gradual and local formations constituting the earth's crust, disclosed by geology, we add the grander theoretical inference, that all the varied modifications of animal life were equally produced according to some regular scheme of physical causes; or, if the more imaginative speculator should think that he can identify that scheme with certain physiological indications of rudimentary evolution,—it is impossible to see in what respect the latter class of views can affect religious considerations *more* than the former, or be *more opposed* to the letter of the Mosaic description than they are. Yet there are those who seem to view these last ideas with more peculiar apprehension. The discrepancy cannot really be greater whether we adopt any physical theory of the mode of origination of successive forms, or whether we reject all such speculation. In a word, those who understand and accept geological truths at all, and admit the palpable contradiction to the Old Testament without prejudice to their faith, cannot with *consistency* make it a ground of objection to any hypotheses of the *nature* of the changes indicated, that they are *contrary to Scripture.* They are in no way *more* so, than *all geology* is.

The idea which is often attached to the word "Creation," as meaning a calling into existence *out of nothing* (as already observed), rests wholly upon certain *metaphysical* arguments * which it is no part of my design to discuss. But with reference to the opinions of those who lay so much stress on the *letter* of the Bible in such points, I would merely observe

Idea of creation out of nothing.

---

* The doctrine of the origination of matter *out of nothing* has been upheld on various metaphysical grounds; but it has received peculiar support from the speculations of Oken, in his work before referred to. (Elements of Physio-Philosophy, by Lorenz Oken, M. D., Prof. Nat. Hist. Zurich. Transl. by G. A. Tulk, M. R. C. S. Published by the Ray Society, 1847.

His whole theory rests, in the first instance, on certain views of what is termed "Mathesis" and "Ontology," which consist in deriving the original existence of matter on the principle of the algebraic formula $0 = + -$, or "out of nothing there is something," which is applied in an hypothesis of antagonistic principles of existence, under certain conditions neutralising each other, but under others evincing independent existence.

From metaphysical principles of this kind the author deduces the existence of an "eternal self-consciousness, which is God" (§ 61.), and even the *mode* of his existence leading to the doctrine of the Trinity (§ 67.).

Again, on similar grounds, introducing the agencies of heat, light, ether, and especially "polarity," from a combination of these with the antagonistic ideas before mentioned, he derives monads, and thus the world's development out of nothing by processes which, he says, are the exact "Genesis of Moses," terminating at length in the creation of man. (§ 958.).

He ascribes the existence of the universe in the first instance to the Divine will: "God spake and it was" (§ 63.) And, again, "God has made heaven and earth out of nothing. God has not found matter co-eternal with Himself, and, like an architect, arranged this to His fancy; but he has out of His own eternal omnipotence by His will simply evolved the world out of nothing into existence." (§ 167.) But all this will not satisfy the bigots, who set down Oken as an atheist!

that, whatever may be the value of other arguments
in support of it, it is wholly destitute of any foundation in *Scriptural* authority. The word which in
Genesis and elsewhere, is rendered " create," has
been pronounced by eminent Hebrew scholars* by
no means to bear the sense above mentioned, being
only a stronger or *more intensitive* form of expression
of the idea of *making* or *fashioning*. While other
passages leave the idea at least equally indefinite;
if, indeed, they do not in some sense refer to pre-
existent matter.†

Organised
beings not
formed out
of nothing.

Moreover, in the particular instances of the creation of animals and plants, it is notorious that in the
very language of the Mosaic narrative, the aquatic
animals are described as being brought forth *out of
the waters* ‡, as the land animals and plants are *out
of the earth*, and man especially, as formed *not* out
of nothing, but out of " the dust of the ground."
Thus, at any rate, those who maintain that the first
individuals of all new species were always brought

---

* On this point the reader is referred to the authority of Dr. Pusey in
Buckland's Bridgewater Treatise, i. 24.
† See *e. g.* Heb. xi. 3.
‡ Gen. i. 11. 20.; ii. 7.

into existence out of nothing, can at least have no
shadow of *Scripture* authority for such a belief.

But the most strange and inconsistent part of the
whole seems to be, that those professing such literal
adherence to the Mosaic narrative should utterly disregard it in having recourse to so many successive
repetitions of the work of creation in different epochs,
when the whole drift and purport of that narrative
is manifestly and palpably directed to the one special
object of representing the whole as a single creative
act, *begun* and *completed* in the six natural days, with
peculiar and emphatic reference to the *final cessation*
and rest on the seventh. Yet more inconsistent are
those who contend for the primeval Sabbath, and yet
uphold the seven periods of unlimited length !

*Inconsist-
ency of re-
peated
creations.*

The prevalent theology is too deeply immersed in
an indiscriminate and unthinking Bibliolatry. But
even on the fullest admission of inspiration, the
slightest rational reflection must show the unreason-
ableness of looking for indications of the inspired
character of Scripture, in relation to any other subjects than those of its proper spiritual communica-
tions ; and even these in the *mode* of their introduction
are always specially *adapted* to the apprehensions and

*Scriptural
representa-
tions
adapted to
the state of
knowledge.*

condition of those to whom they were addressed, and always to be applied subject to the due discrimination of circumstances, times, parties, and dispensations.

Reference to creation in the New Testament.

Thus, more precisely with respect to the subject of " Creation," the writers of the New Testament, doubtless adopting themselves the existing belief respecting it, yet never dwell upon that belief in detail*, nor insist on any of its peculiarities. They refer to it, in fact, only in a general sense as opposing the superstitions of heathenism †, and teaching the Gentiles that the elements of the material world, which, either directly or under various mythical personifications, had been the object of their worship, were, in reality, the *creatures*, not the *Creator*, to whom alone worship was to be given. The only specific references made, are those of a more elevated and mysterious nature, involving *no physical ideas*, but referring the work of creation to the Divine Logos ‡; probably in refutation of the speculations

---

* We may except one solitary instance (an exception which eminently proves the rule), when the Apostle is specially arguing with *the Hebrews*, and, referring to their belief in the Divine rest on the seventh day, applies it figuratively to the future and everlasting rest of the faithful. (Heb. iv. 4.)

† Acts, xvii. 25.

‡ John i. 1.; Col. i. 16.   See Dr. Burton's Bampton Lectures, p. 112.

of the Gnostics; or, as some think, in giving a Christian sense to them.

To the same kind of misapprehension may be traced — but even with less appearance of reason — the zeal with which the belief in man's *recent origin* on the earth has been maintained, and the suspicion and animosity excited by even a hint or conjecture at any possible higher antiquity of the race. The prevalent belief in the very recent origin of man, geologically speaking, depends wholly on negative evidence. And there seems no reason, from any good analogy, why human remains might not be found in deposits corresponding to periods immensely more remote than commonly supposed, when the earth was in all respects equally well suited for human habitation. And if such remains were to occur, it is equally accordant with all analogy to expect that they might be those of an *extinct* and *lower species*. The only real distinction which marks a supposed " human epoch " is not the first introduction of the *animal man* in however high a state of organisation, but the endowment of the animal with the gift of a moral and spiritual nature. It is a perfectly conceivable idea that a lower species of the human

Belief in the recent date of man's origin.

race might have previously existed, destitute of this endowment.

From the Hebrew chronology.

The belief in the recent date of man is usually adopted from the received Hebrew chronology, itself (as is well known) open to critical difficulties. But, indeed, to those who imagine the Bible authoritative in matters of philosophy or chronology, there is no limit to inferences of this kind. There are some, even, who believe that the " permanence of species " is a *Scriptural* doctrine, because it is said that plants " after their kind " " have their seed in themselves ! "

Objection as to the origin of man.

But the idea of a physical process of origination of organic life has excited a more peculiar opposition, on the ground that it would include MAN *and his descent* in the general category, and represent the human race as at some remote period gradually developed out of an inferior species, which, it is alleged, savours of materialism, and lowers the moral dignity of man. Now, agreeably to what was advanced in a former Essay*, it must, I conceive, appear, that in proportion as man's *moral* superiority

Physical origin of man independent of his moral nature.

---

* See Essay I. § II., and Essay II. § II.

is held to consist in attributes *not* of a *material* or corporeal kind, or origin, it can signify little how his *physical* nature may have originated. The same moral superiority may equally belong to him whether originally evolved out of any form of lower organic life, or out of a clod of earth. All truths relative to man's moral or spiritual nature, in proportion as that nature is held to be of an *immaterial* kind, must be allowed to be entirely independent of any theories of the origin of his animal and material existence.

The difficulties felt on this subject by some seem mainly to arise out of the belief as to man's primeval state. But even the Mosaic account, it is admitted by most interpreters, altogether refers, *not* to man's *physical constitution*, but to the peculiar *spiritual* nature given to him; expressly described as " breathed into him " * by a special act, and which is generally conceived by divines to have constituted " the image of God,"† in which he was made; and from which, according to the received view, he fell; all which can surely in no way be affected by what may have been his *animal* nature or origin *prior to*

Objection with regard to the primeval state of man.

---

* Gen. ii. 7.   † Gen. i. 27.

K K

that spiritual creation; as it refers to that part of his nature which is spoken of expressly as distinct from, and independent of, his physical constitution and material organisation.

But if we look to the New Testament view of the matter, it will be perceived that the Christian argument assumes man in a state of degradation and sin, from which it would elevate and transform him by the renovating power of divine grace. As to any previous state, or the origin of that depravity, St. Paul, even in adopting the representations of the Old Testament, dwells on no details, but directs the whole stress of his argument, not to the physical *history or origin* of the evil, but to enhancing the greatness of the *deliverance* from it, and points to Adam only to lead men to Christ.*

Does not really affect the doctrine.

Thus the adoption of philosophical views of orderly evolution will not be found to impugn religious doctrine. Thus the theologian can have no ground for denouncing such physical speculations as impious or subversive of scriptural truth. Were theories of development ever so well established,

---

* Rom. v. 20.

they would not affect those doctrines; they do not
even contravene the letter of the physical representations of the Old Testament to as great an
extent as all geology does, and still less do they
offer any opposition at all to the more spiritualised
representations of the New Testament.

To urge objections, however, on theological
grounds against such theories has been a popular
topic with a certain class of writers; and it constitutes
the main object of a work, already referred to, which
has attained a more especial reputation among those
who adopt such theological views as those just glanced
at, or who conceive physical theories necessary for
the support of religious faith, — Mr. Hugh Miller's
" Footprints of a Creator," &c. So far as the
author's observations bear on real points of geology
and palæontology, they are characterised by his well-
known acuteness and power of illustration. Yet
throughout the whole we cannot but observe that
the polemical spirit and avowed theological bias with
which it is written cannot but weaken the authority
of many parts of his physical argument.

The author introduces with great effect remarks
on the high organisation of the early fishes, on which

Objections brought forward by Mr. Hugh Miller.

he is so well qualified to dilate, and which he enlarges upon with so much animation as furnishing indubitable "footprints" of the Creator; — that this is so every reader will most willingly admit, but in what sense it is *more peculiarly so* than is supplied by the contemplation of any other organised structures, or why more so than if a successive development were made out, it is to my apprehension impossible to conceive.

The author, though a very strenuous theological champion, is yet candid enough to allow that the theory of transmutation, if established, would not be absolutely "atheistic;" yet he considers it hardly less destructive to religion, because he conceives (unless I mistake his meaning) it would make the human soul a part of the mere development from the material world; and again, because it is opposed to the doctrine of man's primeval innocence *, though in what way it could affect the former doctrine more than the fact of man's natural birth from a parent does, or (in accordance with what has just been said) how it is at all related to the latter tenet, is equally difficult to perceive.

---

* Footprints, &c., p. 17.

There are many who in their zeal for the authority of the Old Testament overlook, or, indeed, seem altogether ignorant of, the better views disclosed in the New; and it is to this Judaical school that Mr. H. Miller's theology seems traceable, as appears in a more special manner from some remarks towards the conclusion of his volume. But, without diverging into irrelevant particulars, I would only observe, with regard to the pervading principle of his work, that the author appears to consider the entire question as one between the idea of Divine operation, and what by a very common confusion of thought he regards as antagonistic to it, the principle of natural law and order, and censures " those who would transfer the work of creation from the department of miracle to the province of natural law, and would strike down in the process of removal all the old landmarks ethical and religious," * as if the great principle of natural laws and the order of physical causes were not as entirely the emanation of the Supreme Mind, as any supposed intervention could be, and, in fact, the only true proof of it.†

Origin of objections in narrow Judaical views.

---

* Footprints, &c., p. 17.
† I will add the following expressions in a letter from a friend as a

In no way affecting Christianity.

It can only be in the want of a more worthy appreciation of the true nature and real independence of Christianity that we hear denunciations of more enlightened views as " removing the landmarks of all religious and moral obligations," because the authority of the Decalogue and the Sabbath is thus invaded, which, however, a slight attention to the language of the Apostle of the Gentiles must convince any unprejudiced inquirer has been in modern times unhappily mixed up with Christianity *in a way directly opposed to the whole spirit and tenor of his teaching.**

Nor can it fail to be observed how miserably low must be the notions entertained of the grounds of *moral* obligation, if they can be supposed implicated in a question as to the process or order of the *physical*

---

commentary : " Seeing the Creator in His laws seems an idea worthy of a Christian philosopher. A Hindoo wants a god made for him by the priest, before he can understand or will allow the idea of a God : and only such a one should demand in natural theology, the occasional making of a new animal out of mud."

* The practical influence of these Judaical views, displayed in the spirit of Sabbatism, is sometimes unhappily exercised even over science. Humboldt justly satirises " the English Sunday, on which it is sinful after Saturday night at twelve o'clock to read off a scale," as having destroyed the value of an important set of magnetic observations. (Cosmos, 1st transl., note 113. to p. 188.) Yet extensive tables of certain observations are still printed in which every *seventh* entry, instead of degrees, minutes, and seconds, is filled up by the word " Sunday ! " It would be a curious calculation to find the real value of a *mean* deduced from such a column !

On this subject the reader is referred to my " Essay on the Law and the Gospel," Kitto's Journal of Sacred Literature, No. II. April 1848, and to my two Sermons, " Christianity without Judaism," 1856.

creation. Whether those grounds be regarded as connected with intuitive and immutable natural principles, or whether they be referred to the simple authority of Christian precepts, the moral law of the Gospel built upon faith, they must be equally independent of all theories of *creation* and of the *Judaical law.*

In connexion with this subject, one other argument may here be noticed because it has been dwelt upon by some writers of eminence — the application of the supposition of " Successive Creations," in the sense of interruptions, in support of the belief in *miracles* generally; but this argument (apart from the hypothetical nature of the events assumed) will easily be seen to be of very little force, when we recollect that these so-called " creations " were, by supposition, events *constantly recurring, and essentially different in their entire nature and circumstances* from any alleged miracles, wholly unconnected with any *revelation,* and according to the very terms of the assumption, they were the commencement and establishment of a series of *natural* results, of which (according to the view commonly adopted) miracles are professedly the violations.

Successive creations made an argument for miracles.

K K 4

Those, indeed, who think it more satisfactory to adopt that view of miracles, which has obtained the sanction of so many eminent and orthodox divines * — (assuming the question of testimony) that instead of interruptions they are really to be regarded rather as instances of the observance of some more comprehensive laws unknown to us, — will of course see little value in such an argument as that just referred to, but will naturally feel it much more congenial to their ideas to fall in with the more elevated conceptions of law and order presiding over even the earliest changes and evolutions of the organic world.

It is probably in reference to the species of argument just mentioned that Mr. H. Miller expresses his opinion that in the present age "the battle of the evidences will have to be fought on the field of physical science;" † but if it be on the fair field of true inductive philosophy the victory will clearly be on

Does not affect the evidences of Christianity.

---

* This view has also been remarkably elucidated by Mr. Babbage (ninth Bridgewater Treatise, ch. viii.) from the nature of "laws intermitting," as exemplified in several parts of mathematical analysis and in his own calculating engine — that is to say, a mathematical formula, or a series of mechanical movements — is originally so constituted and framed that it shall give a long series of results of one continuous character, but at some one point shall exhibit a singular apparent interruption of that series or deviation from it which is nevertheless really as much a part of the series as any of the more regular terms.

† Footprints, p. 21.

the side of law, design, arrangement, and subordination of causes, as the true exponents of supreme creative power and wisdom, and the real evidences of *natural* theology.

With regard to those of Christianity, in the opinion of some approved divines they mainly rest on the internal and moral proof which it carries with it; and the more the age advances in real enlightenment the more will its purely spiritual claims be evinced, as wholly independent of those adaptations of an earlier dispensation, restricted, as they were suited, to the condition, the ignorance, and wants of a particular people in long-past ages; but which, nevertheless, even at the present day, are still by many strangely regarded as if they were designed for permanent and universal truth.

The Christian doctrines, from their very nature, are conveyed in the language of the spiritual world; they belong altogether to a higher order of things; and where they may be expressed as in any degree related to material objects or events, these representations cannot now be canvassed in detail, nor fall within the province of physical investigation:

what are expressly described as supernatural mysteries, are, as such, exempted from physical difficulties.
The truths they embody shine calmly by their own
heavenly light, like the stars above the brighest il-
luminations on earth.

All external evidences must necessarily vary in
their nature, force, and application, as addressed to
different ages and persons of different capacity;
unless so adapted, they must fail in their object.
And this accords with what we find was the
actual method and practice of the founders of
Christianity in their appeals to the different parties
and classes of mind they addressed.

The evidences of natural theology (such as they
were in that age) are expressly recognised by
the New Testament *; and it is, therefore, in entire
accordance with its spirit that we follow them out
at the present day in any more extended speculations to which we are led by improved science, and
by which we may be able better to elucidate the
order and method of the visible creation.

Natural
theology
progressive
with
science.

Though it is the attribute of Divine truth to be
one and the same for ever, it is no disparagement to

---

* See Rom. i. 20.; Acts xiv. 17.

that invariableness that natural theology should be *progressively changing* in the *aspect* and character of its evidence, with the improvement and advances of those sciences on which it is founded; and thus leading to *more enlarged and worthy conceptions* of the Infinite and Supreme Intelligence, which cannot but exert a beneficial influence on the views subsequently formed of more particular doctrines. Thus to shrink from any investigation because it may seem to disparage hitherto accepted ideas, or to unsettle old convictions, is a mere mark of weakness and timidity on the part of its advocates which is inconsistent with the resolute pursuit of truth, and can end in nothing but endangering the very cause they seek to serve, and yielding up the vantage ground to their opponents.

To recapitulate and conclude: as in the *existing* condition of the material world, in those phenomena which are best understood and most perfectly investigated in all their laws and relations, it is that we have the highest and most indisputable evidences of the Supreme Moral Cause; so in regard to the *past* in the same way, where we can best trace the steps and processes by which the changes have gone on,

Conclusion.

Evidence of creation not in interruption but in orderly processes.

there we recognise the true evidences of creation. Yet it is the very reverse of this view which a certain class of writers would seem to uphold. They would seek the proofs of creation, not in the *known*, but in the *unknown*, regions of Nature; and precisely in those instances where we are *least* able to trace order and system in the Divine design, there they think we should *most properly* find its evidence! that we should acknowledge its proofs rather in the *ignorance* than in the *knowledge* of those recondite laws by which its reason is manifested! that we should behold the Deity *more* clearly in the *dark* than in the *light;* —in confusion, interruption, and catastrophe, *more* than in order, continuity, and progress.

Parallel in human governments.

If in travelling in a strange country we see around us the signs of order and security, civilisation and improvement, law and justice, maintained without the violent or visible interference of authority or force, we immediately infer that we are in the territory of a firmly-established, wise, and beneficent government. But if we pass into another district, where all these signs are wanting,—where we find the country in a convulsed and tumultuous state, the social machine unhinged, and the arm of power displayed only in coercion,—if we should be told that this district was subject to the same government, we might reasonably doubt or deny it; assuredly we should not infer it. But most certainly, if the lawless and convulsed state of the country and the display of arbitrary force were to be pointed out to us as the *special proofs and indications* of such dominion; — if our guide were to say, " Behold here the proper display of the majesty of the law, — here the true evidence of the greatness and wisdom of the ruler, of which you see nothing in the peaceful territory,"—we could only regard it as a mockery. Yet such is the argument of " The Footprints," and other popular works of the same class.

Imagined interruptions of preordained order for the introduction of new forms of life, so far from evincing perfections, must appear rather like blemishes in the beauty of creation; marring the picture by blots or blanks, where we should fail to follow the outline or trace the artist's design, — however we might, from the surrounding parts, conjecture its continuity.

Interruptions of law interruptions of evidence.

If, in following the track of a person, we for a time lose sight of it in broken ground, and after-

Evidence of footprints.

wards regain it, it would be absurd to say that we *discover his footsteps* in the broken ground, because we may *infer* that he must have passed over it; so we see the rational evidence of creative power and wisdom wherever we can trace the particular steps, laws, and processes by which its operations have proceeded; but where the order of causes may be as yet hidden from us, to say that there *especially* we recognise the "indications" and "footprints of the Creator" is a contradiction. These are exactly the points in which those indications are *wanting*.

Wanting only in the origination of new species.

Through all past time we discern *everywhere* the footsteps of the Creator,—in all extinct as well as all living organic structures, modelled upon one plan,—in their marvellous affinities, all linked in one chain,—in the whole scheme of one continuous series of causes in which they are united,—everywhere, *except only* in the *mode* of their *origination*, because that is not as yet traced to its law; *there alone*—these *footsteps* are, in consequence, at present *concealed* from us, though even there analogy points to them through the principle of orderly evolution.

Science demonstrates incessant past changes, and dimly points to yet earlier links in a more vast series of development of material existence; but the idea of a *beginning*, or of *creation*, in the sense of the original operation of the Divine volition to constitute nature and matter, is beyond the province of physical *philosophy*; it can only belong to that of *faith*, and find expression in the language of *inspiration*.

The beginning a matter of revelation.

But though we know not what was the beginning, or will be the end, of created things, and though the whole of the present, equally with the past, be but changing phases of existence, and the material universe itself be but perishable and transitory,—yet LAW and ORDER existed before them, and will continue after them; HARMONY and SYMMETRY are permanent and eternal,—the archetypes of the Divine plan,—the very impress of that Supreme REASON and " WISDOM,"* which " was set up from everlasting;" which " when He prepared the heavens was there," and " when He appointed the foundations of the earth was by Him;" the Divine LOGOS of the

Order the Divine archetype.

---

* Prov. viii. 23. 29.

Christian Genesis *, who "was in the beginning with God, and who was God: by whom all things were made, and without Him was not anything made that was made."

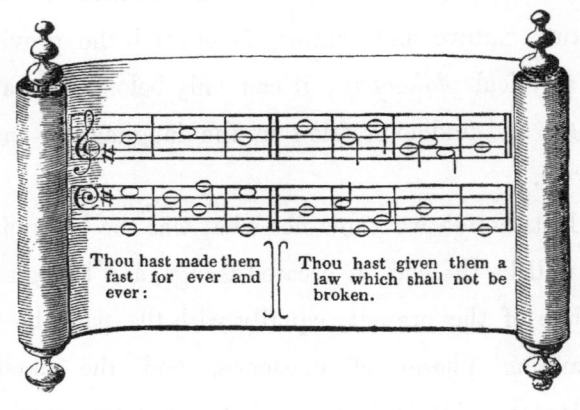

Thou hast made them fast for ever and ever: Thou hast given them a law which shall not be broken.

* John i. 1.

APPENDIX.

APPENDIX

APPENDIX.

No. I.

On Inductive Reasoning.

Note to pp. 6. 13.

THE well-known passage in which Aristotle * analyses the logical nature of inductive proof, and its relation to syllogism, is confessedly not only obscure, but seems to involve a contradiction, especially in his contrasting syllogism and induction in one place, and yet showing that induction may be reduced to syllogism in another. The difficulties of the case have been elaborately discussed by Dr. Whewell.† Aristotle's view is simply reducible to this, — that when an inductive argument is put into the form of a syllogism, it necessarily involves, as the major premise, the assumption that *all* objects of the kind, of which *some* are enumerated are *like those enumerated* in the particular respect specified. With this assumption formally introduced, the syllogism is perfect in point of form. This is exactly what is pointed out by Archbishop Whately.‡ Or, in other words, it amounts to

---

* Anal. Prior. ii. 25. † Camb. Trans. ix. pl. 1.
‡ Logic, bk. iv. ch. i. § 1.

saying that the argument of induction cannot be reduced to a syllogistic form except by formally making this assumption.

The only real question involved is, as to the means of arriving at the *truth* of this assumption. But it is a main consideration often overlooked, that the *truth* of the *premises,* or the source whence we derive them, is, in the language of logicians, a matter wholly " *extra dictionem,*" and with which the syllogistic theory, as such, is no way concerned. But if we proceed to consider the origin of this assumption, it is no doubt arrived at by a process of reasoning and abstraction. Aristotle says it is necessary νοεῖν — to conceive it by an intellectual act. The question is as to the nature of this act; and this is what I have above endeavoured to elucidate.*

Mr. Mill questions the principle of this assumption, and, in some respects, opposes the views of Archbishop Whately †, yet seems to admit that every induction is or is not valid according as the particular instances adduced are or are not sufficient to make it allowable to draw the general inference. But it may be asked whether to decide this in the affirmative is not, in fact, equivalent to making the assumption in question ?

It is also important to bear in mind another distinction (which Mr. Mill himself has elsewhere admirably illustrated and insisted on ‡), viz., that, between the *de facto* origin and sources of our convictions, on the one hand, and the *logical order* into which they *may be analysed*

---

* On this subject the reader is referred to an acute discussion in a small work entitled " A Delineation of the Primary Principles of Reasoning," by the Rev. R. B. Kidd. London, 1856, p. 256.
  † Logic, i. 373.   ‡ Ib. i. 267.

on the other, they differ as a physician's prescription
differs from a chemical analysis of the ingredients.

With respect to the peculiar "inductive principle"
maintained by some, it is alleged that we have a certain
inherent principle of knowledge, which nevertheless requires for its operation the exhibition of a certain amount
of external facts ; and this has been eloquently compared
to the inherent powers of vegetable life, as in a bud, to
develop to a flower; yet the external influences of sun,
rain, &c., are not less necessary to its action. Or, again,
in the same way, the eye, constructed as it is, could not
see without light ; or, otherwise constructed, could not
see in the light.*

But all these illustrations, apt and imposing as they
are, are after all of little real force, if the first assumption, that we *have* such a distinct internal power, is
shown to be unnecessary, and that there is nothing
really peculiar in the case, as has been attempted in my
first Essay, § I.

The general assumption that the mind has a power of
inferring with certainty more than actual experience
warrants, presented in a variety of forms of illustration,
forms the substance of various speculations on this question; all which, I believe, are capable of analysis, and
therefore ought to be subjected to it, up to simpler principles. The favourite practice is to avoid this labour, by
setting down everything as a *peculiar ultimate principle.*

Thus, Descartes speaks of the " seeds of truth which

---

* See an able article in the Edinb. Review, Jan. 1852, p. 28. ; also
De Morgan, Formal Logic, p. 32.

exist naturally in our souls. " *   Others have contended for innate " capacities," not ideas, and have asserted a combined action of these capacities with the experience of the senses, as leading to inductive generalisation.

A proof of such intuition is often alleged in mathematical axioms, or in the deduction of *necessary* truth generally, on which I have commented at large in my " Essay on Contingent and Necessary Truth," before cited, and I conceive have indicated that they *may be* reduced to simpler elements requiring no such theoretical assumption.

In those parts of these Essays which bear on metaphysical questions, I have adverted little, if at all, to their degree of accordance or discordance with those of the most celebrated metaphysical writers.   This has arisen chiefly from the wish to avoid going into controversy, or appearing to advocate the tenets of any particular school.   But a few remarks may seem called for by the importance of some of the topics adverted to, in extension and continuation of those above offered.

In what I have advanced, more especially on the " inductive principle," and on the nature of " causation " (differing from the views often maintained, at which I have glanced), it will be seen by readers versed in the writings of KANT, that some of the topics discussed run very nearly parallel with those of his celebrated investigations. It may therefore be advisable to add a remark or two, as to the degree in which my ideas may seem to resemble, or to be opposed to, those of so eminent a philosopher.

* Méthode, p. 5.

In general, as to the physical sciences, Kant observes most truly, that when they first assumed a truly philosophic form, in the investigations of Galileo, it became clear that man is not the *passive disciple*, but the *judge* of nature," * starting *à priori* physical problems ; that such principles of reason exist in nature, and thus physical science and induction are not mere empiricism, but founded in *reason*.

I need hardly remark how exactly this agrees with the eloquent observation of Œrsted quoted in my first Essay, § I., or how fully I concur in, and have endeavoured to analyse and illustrate, the same truth. The "principles of reason" which "exist in nature" are undeniably brought out by our abstractions, whence we can reason downwards, and in many cases *predict* physical results ; but there must be an original abstraction from experience, to lead us to those natural principles.

The question of the origin of our knowlege, and how far it is or is not entirely derived from experience, as is well known, is largely discussed by Kant, and forms, in fact, the basis of his researches.

When, in the introductory part of his work, he asserts that "no knowledge precedes experience, all commences with it," he yet draws the distinction between "commencing with experience" and "coming from it," which he illustrates by the example, "Every murder supposes a murderer." Here experience furnishes one element, the *matter* of knowledge — the ideas of a murder, of a murderer ; but the other element, the *formal* part (as he

---

* See Cousin's Lectures on Kant, transl., p. 19.

calls it), is equivalent to the assertion "Every change supposes a cause of change," which, he says, surpasses experience, yet could not commence without it. This
element he considers to be derived from the mind itself,
or to arise à *priori.* "Every event *must* have a cause,"
is a maxim which "anticipates all future experience, and
is independent of all past experience," though even here
one element, "change," is derived from experience, but
the other, " necessity," is not.*

Nothing, I think, can be more clear or masterly than
Kant's reduction of the question to the wider and essential point of the difference between *contingent* and
necessary truth : and it seems in complete accordance
with his own views if we advance one step further,
and remark the correspondence of this distinction between that of " experience," which is "contingent," and
the logical deduction of one truth *from another*, which is
"necessary ; " and this I believe constitutes the entire
idea of "necessity," — *necessity of reason*, which is wholly
relative to some previously established truth.

We may, if we please, *analyse* the inference (in the
case supposed) up to the maxim " Every event must have
a cause," though we do not in practice so deduce it.

But taking the theoretical analysis (observing that
the word "cause," as here used, involves no opinion as to
the abstract nature of causation,) it refers to a mere
inference from the general observation of *changes*, that
they are connected in a series, which is a generalisation
from experience, and "surpasses experience " only in
the same sense as all inductive conclusions do.

Or we may, if we please, set out from some higher

---

* See Cousin's Lectures on Kant, transl., p. 23.

abstraction from experience, the idea of "an event" in general. When we have formed this abstraction, and defined it at our pleasure, we may reason upon the definition so formed, and may come to the conclusion that every "event," in the sense assigned, must be preceded by another. Whatever may be the nature of this definition and the reasoning upon it (supposing it logically correct), the conclusion will have the same degree of certainty, and no more, with that of the original abstraction of the idea of an "event."

Kant, when he proceeds to the more full analysis of the question (as is well known), recognises two sources from which all knowledge is derived: (1.) the "sensory," which is merely recipient and passive, and gives representations or, what Cousin renders "intuitions" of phenomena; (2.) what is rendered "understanding," but seems to me to correspond to Locke's "abstraction," which is active, and forms "conceptions" or "notions," and "spontaneously developes itself." — The study of mental operations belonging to the two respectively, and hence termed "Æsthetic" and "Logic."

It is, I apprehend, to the closer analysis of the 2nd faculty, corresponding to "abstraction," that the solution of the whole difficulty of what has been termed the "fundamental antithesis" of all philosophy — that between sensation and idealisation — may be referred.

Any seemingly preliminary general assumption, such as "Every event must have a cause," is no really *à priori* idea. In the *first* processes of induction we *do not* adopt any such generalisation; it is the result of the exercise of abstraction and comparison, by which we gradually and insensibly come to anticipate the senses, and often, indeed almost always at first, *erroneously.* In a word, "necessary truth" is nothing but "necessary consequence;" and this is purely a matter of logic — the result of a series of abstractions combined together. That the general must include the particular, is merely an essential part of the idea and process of abstraction.

Experience collects particulars; abstraction or generalisation, so far from *adding* anything which experience does not supply, in fact *takes away* all the points of difference, and leaves only the points of resemblance as the naked abstraction or *general* idea. It is only with ideas thus abstracted that we can reason.

The ideas of time and space are, I believe, not formed at all till after long experience; they then result as highly generalised abstractions. Consciousness is, I conceive, nothing else than an abstraction from continual, universal experience. Kant connects the idea of time with that of consciousness, as on this view would be the case.

On these grounds I agree with him, that " a complete analysis shows that every thought can be directly or indirectly traced to the 'intuitions,' and consequently to the sensory." *

As to the idea of space, he observes, " How can there be in the mind, before any objects whatever have been presented to us, an internal intuition which shall determine the conception of such objects ? It must be that it exists in the subject as a formal capacity of being affected by objects, and of receiving from them by this means an immediate representation, that is to say an 'intuition,' —a form of the external sense." †

Nothing can be a more distinct exposure of the notion of ideas or truths existing originally in the mind, though perhaps the phraseology employed is not so clear as might be wished. Grounded, however, on these distinctions, Kant classifies all sciences under the two heads of those derived à *posteriori* or empirically from experience, such as the natural and experimental branches; and those investigated à *priori*, as arithmetic, geometry, &c. In particular, he contends that the mathematical sciences involve the principle of identity, but carefully distinguishes that they do not originate out of it; it is only a necessary *condition* of their deduction.*

When Kant observes that the idea of "space exists in the subject as a formal capacity," &c., it is surely saying nothing more than that the power of abstraction exists in us, by which we form such a representation or idea.

Reid and others have asserted "innate capacities," instead of innate ideas, which seems to me to add very little to the explanation of the case. I believe that all supposed à *priori* principles are really reducible to the results of what Kant recognises as the *active* principle (abstraction), superadded to the "sensory" or *receptive* to which last they must, however remotely and indirectly, trace their first origin.

The distinction between science observational and abstract, à *posteriori* and à *priori*, I believe to be merely one of *degree:* in the lowest collection of familiar facts there must be idealisation; in the highest deduction from a first principle, that first principle is a result of generalisation—of abstraction from particulars originally ac-

* Cousin's Lectures on Kant, transl., p. 32.

quired by observation and experience external or internal.
The reasoning downwards is *necessary* reasoning ; and
thus the result is *relatively* a necessary truth. In many
instances we can arrive by two or more such courses at
the same result : where we follow deduction from a
highly simple first principle, there we have the more
purely necessary positive science ; where from a lower,
or only from the lowest, there the less pure and necessary,
and the more contingent.

This is what I have endeavoured to exemplify in several
parts of these essays, especially Essay I. § i. pp. 21. *et seq.*

As to the principle of identity in mathematics, I have
elsewhere, as I conceive, shown that identity of quantity
is not the essential idea, but difference of operation on the
same quantity,—that the general expression, in a word, is
not the celebrated $a = a$, but $f(a) = \phi(a)$ where the functions or operations designated by f and ϕ are essentially
different.

V. Cousin, in commenting on those portions of Kant's
principles just referred to, does not conceal his hostility
to the "sensational" school ; and is not sparing in
animadversions on Kant whenever he discovers in him a
leaning towards its doctrines, or in expressions of triumph
when he thinks he can adduce Kant as an auxiliary
against it.

Cousin insists mainly on the assertion that all induction rests on the primary assumption of the stability of
the laws of nature, which he says is beyond all experience,
and attacks the sensational philosophy as inconsistent in
admitting, as he conceives it must do, this principle,
observing that " the necessary cannot follow from the

contingent." And here he considers Kant's views as tending directly to the complete overthrow of sensationalism, in admitting this first principle.*

He also objects to Kant's making consciousness depend on the sensory, as supposing it, thus, passive and receptive, which he regards as absurd and contradictory.

As to causation, Kant, though generally upholding Hume, yet censures him for decrying *à priori* principles as fancies of the imagination and originating in nothing more than a habit explicable by experience and its laws, and thus purely empirical, and in no way characterised by necessity or universality. " To support this novel opinion," Kant says, " he appeals to the commonly adopted idea of the relation between causes and effects," and infers that there is no real necessary relation on *à priori* grounds. Thus, *e. g.,* no *à priori* principle could teach us that the sun's rays would melt wax and harden clay, whereas Kant contends that we can infer *à priori* that something has preceded the facts in question, and that they are due to some constant law, though it is for experience to determine *what* law. Hume, he contends, erroneously concludes the contingency of the law itself from the contingency of the actual cases of its application, and hence reduces the principle of causality to mere association of ideas and contingent relations.†

That we can and do infer that all phenomena are due to some constant law, that everything in gradation may be traced to successively higher principles, as Kant most truly affirms, is precisely what I have here contended ; and this is what I conceive must be super-

---

* Cousin's Lectures on Kant, transl., p. 32.   ‡ Ib. p. 152.

added to Hume's principle of mere "invariable reference," in order to give a just philosophical view of the nature of causation : but still all this is purely the result of successive *abstraction.*

The objections thus referred to seem to me to be comprehended in a very small compass, and to be all of a kind to which the remarks here advanced on the inductive principle furnish a sufficient reply. We *do not,* I apprehend, even pretend to "derive the necessary from the contingent." The *necessity* of any conclusion relatively to the premises, is merely a part and consequence of the nature of abstraction ; and there is no other *necessity* in any truth. The very highest abstractions are only results of experience, co-extensive perhaps with human thought, and hardly separable from our nature.

"That all the phenomena of nature are referrible to some constant laws," is an universal truth, no further *necessary* than the necessity of reason makes it. Perpetual extensions of the principle of natural law, more and more comprehensive, are being constantly worked out, often only in abstract theory, which may perhaps long wait before it receives confirmation from observation. These necessary deductions, unlimited by any material boundaries, all, however, set out from some truth of experience, however remote and simple, and cannot be more necessary or certain than *it* is.

In connexion with the same topics a recent small publication, "An Inquiry into Speculative Philosophy," &c., by A. Vera, late Professor in the University of France (1856), demands a brief notice.

The disparagement of Bacon with which the author

commences, in fact, turns upon denying what no one does or can assert; viz. that he was the *inventor* of the inductive method. It may even be true that he did not essentially *improve* upon its principles. But surely praise enough remains to him, to have been the first to assert and apply it as a philosophical method, to the subversion of the then prevalent scholastic systems, as well as to point out the systematic course which must be pursued in its actual application to the extension of physical discovery.

With respect to the nature of the inductive method itself, M. Vera dwells, with the same emphasis as so many preceding writers have done, on the primary difficulty of the *source* of inductive generalisation. I venture to think that if he had bestowed attention on my first Essay, § 1., he would at least have acknowledged it as offering an attempt to explain that difficulty on principles strictly accordant with a sound analysis of mental processes without assuming any peculiar *à priori* principles whatever, and which, if insufficient, should at least be *shown* to be so.

That induction, even in its lowest stage and degree — the mere collection of facts — implies *ideas*, not mere *sensations*, I quite agree with the author in asserting. No act of observation of the senses is complete, or capable of any application — in fact, cannot be said to be accomplished at all — without idealisation.

If writers on induction, proceeding on what is improperly called the " sensational theory," have omitted to state this, or have currently used language which might seem to imply the contrary — this is doubtless a fault, — yet I believe that generally, if not expressed, this meaning is always *understood*.

But to suppose generalised ideas as previously existing in the mind, and the like, appears to me to involve the too hasty and needless assumption of a gratituous hypothetical principle, whereas I conceive I have shown, in Essay I. § I., that the whole is resolvable into simpler elements.

The author's disparagement of mathematical and physical science (pp. 22. 65.), if understood as referring to the want of clear metaphysical views, in the establishment of first principles and their methods of reasoning, evinced by too many elementary writers, I freely confess, has much foundation ; this evil, indeed, I have myself endeavoured to expose and, I trust, in some parts of the subject to remedy, — especially in several papers in the Memoirs of the Oxford Ashmolean Society. Yet the singular way in which the author himself represents several points of mathematical science, *e. g.* as to curvature (p. 66.), central and tangential forces (p. 24.), the pendulum (p. 25.), &c., seems to imply misconception of the nature of the case of a kind very similar to those of his master Hegel, which have received so full a refutation from Dr. Whewell (Cambridge Transactions, 1849.)

M. Vera thinks that pure induction can constitute no real science properly so called, physical or metaphysical (pp. 18. 21.), since its highest principles and generalisations are professedly derived from the same origin of experience (differing only in extent and degree) as the lowest collection of sensible facts. Hence the boasted claim of physical science to a *high,* and even to the only, *positive* scientific character, he conceives, must fall to the ground. I trust, however, that the view taken in the

1st Essay may suffice to relieve inductive science from this charge.

No. II.

On the question to which I have referred *, respecting the supposed peculiar vital principle, I have great satisfaction in referring to various productions of Dr. Carpenter, especially his article " Life," in the Cyclopædia of Anatomy and Physiology ; his Essay on the Mutual Relations of the Vital and Physical Forces (Philosophical Transactions, 1850), as well as to his Principles of General and Comparative Physiology, 3rd Edit. Chap. iii. Dr. Carpenter has not only shown that the principle of the " Correlation of Forces" may be applied to those concerned in the production of *vital* phenomena, but has recently argued for its extension to *mental* operations in so far as these take place automatically, *i. e.* independently of the will. See his Principles of Human Physiology, 5th Edit., Chap. xi. Sect. 6.

In a communication with which Dr. Carpenter has favoured me, it appears that he views the relation between the mental and physical nature of man, as much more close and intimate than I have represented it in the passages here referred to (pp. 76. 258.), and thus apprehends that we differ much on that point. But in fact I do not at all insist on the *degree* of such connexion or relation. I have only contended that, to whatever extent it be supposed, it in no way affects the moral and religious view of the subject, which rests wholly on *other*

---

* In Essay I. § II. p. 67.

evidence, and refers to considerations wholly distinct *in kind*. To this effect I have now added a paragraph in p. 77.

No. III.

Note to p. 97.

The anomaly of retrograde motion presented by the satellites of Uranus, has been very recently shown to extend to the satellite of Neptune, though much less highly inclined, from the observations of Mr. Lassell, as discussed by Mr. Hind.* Thus the *anomaly* is likely to cease to be one, and to become a part of some greater law affecting in this manner the *outer* planets of our system ; and it would seem to point to some cause acting exteriorly to our system while yet in a nebulous state.

It is also conceivable that the motion of the solar system through space may bring the component bodies of it into contact with other portions of cosmical matter ; as indeed was suggested by an eminent continental astronomer, as the means by which new comets are continually brought within the range of our sun's attraction, and ultimately fixed in our system.

In a valuable paper " on Periodical Meteors, &c.," by Sears C. Walker (Trans. of American Philos. Soc. 1841, vol. viii., new series, Pl. I. p. 113.), a similar suggestion is thrown out, that by the motion of the solar system through space new unformed sidereal matter may be continually attracted into it, under certain conditions forming comets, under others meteorites, showers of shooting stars, &c.

---

* Astron. Society's Notices, vol. xv. p. 46.

No. IV.

Psychology.

The interesting little volume of Psychological Enquiries recently published by Sir B. Brodie throws valuable light on many of the topics noticed in these Essays. I regret not to have seen it till a great part of this volume was printed.

The highly curious subject of the connexion of our physical and mental constitution, which forms a main topic, and is so copiously illustrated through the whole series of these "Enquiries," will throw much light on the points hinted at above.* And the remarks on the comparative endowments of man and inferior animals, full of the most profound interest, will elucidate many of the questions here referred to; but especially the facts mentioned † may have an important bearing on what is here hypothetically suggested ‡ on the relation of man to the system of nature.

No. V.

Note to p. 139.

In addition to what was observed before on causation, it is somewhat curious to notice that, on the other hand, D. Stewart § has adduced this very doctrine of mere observed sequence (discarding the notion of *necessity*), as furnishing the most effectual reply to Spinoza's mate-

---

* Essay I. p. 77.   † Enquiries, pp. 172—179.
‡ Essay II. p. 258.   § Prelim. Diss. p. 110.

rial and atheistic theory derived from the supposed NECESSARY *connexion* of causes and effects throughout nature.

"Necessity" is evidently here spoken of in the confused and mystified sense once adopted, of something inherently *fated* and independent of arranged order, reason, or moral causation. Hobbes's theory of religion is pervaded by the same confusion of ideas with respect to the word "cause," applied indiscriminately to physical causes and moral, above dwelt upon.

No. VI.

Abstract of Professor Owen's View of Vertebral Structure and its Archetype.

Referred to, p. 387.

The investigations of Professor Owen, especially as delivered in his essay "On Limbs," referred to in the text, are so important, that it may be highly desirable to subjoin a somewhat more detailed analysis; in drawing up which, it will be no small recommendation to state, I have had the benefit of the author's own revision and remarks.*

On a cursory view, the skeleton (especially in the higher animals) appears to consist of a chain of vertebræ, terminated by the tail or sacrum at one end and the cranium at the other, while to a portion of the vertebræ are attached ribs, to the sacrum the pelvis, and to it the

---

* See also the same author's Lectures on Vertebrate Animals, 1846; on the Archetype and Homologies, &c., 1848.

lower extremities, and (apparently without any connexion with vertebræ) to the upper ribs the scapula and clavicle, with the anterior extremities.

In the attempt to reduce all these parts to one principle of analogy, Cuvier, Carus, and others, made some advances. The bones of the sacrum were shown to be properly included in the class of vertebræ. But the most remarkable was the idea of reduction of the bones of the cranium under the same analogy, proposed at first merely as an hypothesis, by Oken.

Again, the relations of the pelvis and its limbs, and especially of the scapula and fore extremities, were still not included in the generalisation. They had been referred to imperfect, or even positively incorrect analogies, as being liberated ribs, &c., or even expressly set down as anomalies, by Cuvier, Carus, Geoffroy, and others. * The obscurity chiefly arose from studying too exclusively the higher types, whence the nomenclature was formed on too limited a basis.† The comparative anatomy of lower forms suggests the true analogy.

Now, as to the cranium, had the idea of Oken been supported by the requisite *proofs*, the whole vertebral column would thus have been included in the same analogy ; but, being hypothetical only, Oken's views were opposed by Cuvier and Agassiz, and had become virtually excluded from anatomical science at the period of the communication of Professor Owen's Report on the Homologies of the Vertebrate Skeleton to the meeting of the British Association at Southampton, in 1846. In this the generalisation was revived and established.

* On Limbs, pp. 31. 41. 53. 102.   † Ib. pp. 55. 115.

Again as to the vertebræ and limbs, Professor Owen cleared up the difficulty by commencing with a more accurate view of the *nature of a vertebra*, as a segment consisting essentially of a "centrum"* from which certain "apophyses"† radiate; on one side uniting to form the channel through which the nervous system of the spine is conveyed (thence called the "neural arch," or "neural apophysis"); on the other, the usually larger arch which includes the blood system, viscera, &c., thence called the "hæmal arch," or "hæmal apophysis." These apophyses in some vertebræ take the form of *ribs*, and are here termed the "costal arch." But in different vertebræ these apophyses are differently developed, in some instances being only rudimentary, or having only one or two parts more developed, according to the position and organisation of the part.

But the most essential point (which could never be discovered but by the comparative anatomy of lower forms, and by tracing the development of the higher) is that the arches are *often displaced* ‡ from their vertebræ—sometimes to a greater, sometimes to a less extent; and that to certain arches are attached appendages which diverge from them.§

In this way the bones of the pelvis and those of the posterior extremities are shown to be the developed hæmal arches and appendages of the vertebræ of the sacrum. In the lower forms (as in fishes and serpents)

---

* On Limbs, pp. 43. 81.
† Professor Owen restricts the term "appendage" to the part articulated to and diverging from the "apophyses," whether *neur-*, *pleur-*, or *hæm-apophyses*.
‡ Ib. pp. 50. 61.   § Ib. pp. 78. 105. 116.

the rudiments of those extremities are found, but un-
attached to their proper vertebra or segment. In other
cases they approach in different degrees towards the
condition of attachment and full development.*

The *occipital vertebræ* are the only ones which in the
higher forms appear destitute of a costal arch and appendages *in situ ;* but in lower forms (as in fishes, and
especially in the Lepidosiren) the arch *is seen to be
formed by the scapula and clavicle,* which arch is more
displaced in the crocodile, and still more in the mammalia ;
but the true analogy is thus seen. *The scapula, clavicle,
and fore limbs are the hæmal arch and appendages of the
occipital vertebræ* †, but differently displaced and developed
in different orders by adaptive power.

The same difference in the development and displace-
ment of limbs, according to this analogy, are also shown
in the stages of the fœtal growth in the higher classes. ‡

The undeveloped appendages of other vertebræ are
potential or rudimentary limbs, of which examples are
found in fishes.§

Thus the whole skeleton is reduced into one single
scheme or archetype most resembling the fish form. In
different instances the parts are differently modified, but
always in accordance with one invariable type or system.

As to the insufficiency of the narrow view of final
causes, several striking instances are adduced.

To take a single instance, nothing can be more at
variance with the doctrine that organs are constituted
merely with reference to the purpose they are to answer,

---

* On Limbs, p. 58.   † Ib. p. 69.   ‡ Ib. p. 99.
§ Ib. pp. 60. 65.

than the fact that the bones which in the human hand and arm have their extended development * nevertheless exist in precisely the same number and arrangement, though altered in form, "buried up to the claws in a sheath of tough skin " in the " short trowel of the mole," and " hidden beneath the common undivided sheath of the fin of the dugong or whale." This " offers, perhaps, the most striking and suggestive instance of an *adherence to type, necessitated,* as it would seem, notwithstanding the absence of all those movements and appliances of the limb that explain the presence of the several segments, on the principle of final causes, in the horse and in man."

In like manner, the peculiar jointed arrangement of the bones of the great toe, suited to the purpose of a fulcrum, is strictly preserved in the bones of the foot of the elephant, though all enclosed in one massive hoof, and in the webbed hind-paddle of the seal.

" I think it will be obvious," the author observes, " that the principle of final adaptation fails to satisfy all the conditions of the problem.

" A final purpose is indeed readily perceived and admitted in regard to the multiplied points of ossification in the skull of the human fœtus and their relation to safe parturition. But when we find that the same ossific centres are established, and in similar order, in the skull of the embryo kangaroo, which is born when an inch in length, and in that of the callow bird that breaks the brittle egg, we feel the truth of Bacon's comparison of final causes to the Vestal Virgins."†

---

* On Limbs, pp. 13, 14.　　　† Ib. p. 39.

No. VII.

On the Theory of Unity of Composition, by T. H. Huxley, Esq.,
F.R.S., &c.

Referred to, p. 391.

" In order to a strict classification of animated forms, we must observe that living beings not only *are*, but they *become;* not only have they a definite structure in their adult condition, but each takes˙a definite road — passes through a definite succession of stages — in attaining that condition.

" It is therefore clear that the naturalist must not only make out the resemblance of their *adult structure*, but also the resemblance, in *nature and order, of the successive stages through which they pass.*

" For it is obvious that two living beings *might* have a similar structure in their adult condition, and yet have passed through different stages of development in attaining that condition ; so that the naturalist who classed them together on the ground of their adult condition alone might be altogether wrong.

" To take an example : —

" An error in classification of this kind was made by Cuvier himself. The Cirripedes, or Barnacles, are creatures which in their adult condition present a certain resemblance to Mollusks in many of their structural characters. Cuvier, who knew them in this condition only, did not hesitate to classify them with the Mollusca.

" Later investigators, however, who have studied the entire development of the Barnacles from their youngest state upwards, have shown that at first they are entirely similar to the Waterfleas and Monoculi of our ponds and ditches, which are Annulose animals; and that all the features in which they resemble Mollusks arise from subsequent modifications of their mere external form.

" There is no doubt, therefore, in the mind of any anatomist of the present day, that the Cirripedes are *Annulose* and not *Molluscous* animals. The error of the anatomical method has been corrected by the application of the developmental method.

" In the main, however, and perhaps invariably, when sufficient care has been exercised, the anatomical and developmental methods furnish perfectly harmonious results. Animals possessing similar adult structure, as a rule, pass through similar stages of development; and therefore it has been found that those grand generalisations of purely anatomical facts — upon which Cuvier founded his quadripartite division of the animal kingdom —have been, eventually, only confirmed and placed upon an irrefragable basis by inquiries into development.

" It cannot be too forcibly borne in mind, in estimating the value of Cuvier's system, that he aimed not at a morphology, but at a classification; he did not attempt to discover upon what plans animals are constructed, but to ascertain in what manner the facts of animal organisation could be thrown into the fewest possible genera propositions. He set himself to find out what structural resemblances were the marks of the greatest possible number of other structural resemblances; and having found that the similarity of the structure of the nervous system was a mark of more resemblance of other kinds than any other sort of similarity, he adopted it as the base of his great divisions.

"By such a method, without the study of development, you may have a *classification* of animals, but no *morphology;* you have *without development* no criterion of the truth or falsehood of any doctrine regarding a common plan or archetype.

" What Von Bär did was to generalise the facts of development in precisely the same way as Cuvier had generalised the facts of structure, and to demonstrate that the classification of Cuvier was in the main simply the expression of the fact that there are certain Common *Plans of Development* in the animal kingdom — that there is one Common Plan followed by all Vertebrate animals, another by all Mollusca, a third by all Annulosa, and a forth by all Radiata.

"Finally, the grandest law of all at which Von Bär arrived was, that although beyond *a certain period* in its existence every Vertebrate, Mollusk, Annulose, and Radiate animal followed its own special plan, yet that *up to that point it followed a plan common to all animals;* and thus he demonstrated, and placed upon a footing as secure as that of the law of gravitation, that doctrine of the unity of organisation of all animals which was with Geoffroy an undemonstrated hypothesis.

"In a word, the leading idea on which the doctrine of a 'common plan' now rests is the possibility of demonstrating a *common mode of development* for those animals which are affirmed to be organised upon a Common Plan: — *Development is the Criterion of Homology.*"

No. VIII.

*Abstract of the Theory of Specific Centres.   By the late Professor
E. Forbes, F.R.S., &c.*

Referred to, p. 420.

" 1. A ' specific centre ' is an area occupied by the
individuals of a species.

" 2. It is an ascertained fact, that numerous well-marked
provinces of the earth and sea can be indicated, each characterised by a flora and fauna on the whole peculiar to
itself.

" 3. A *species* absolutely peculiar to a *province* has
necessarily its centre within it; but many species are
common to two or more provinces.

" 4. As a rule, when a species is common to two or
more provinces, these provinces are contiguous, and,
consequently, the specific area is continuous.

" 5. But there are exceptions, such as species, or a group
of species, exhibiting in some cases the phenomenon of
occupying more than one area, or of presenting outliers
of individuals separated from the main assemblage.

" 6. But when we sift the history of such exceptions,
we find that, by tracing back the history of the distribution of the species, or group of species, so situated, in
time (*i. e.* their geological history), we can show the
strong probability of an epoch when all the individuals
of the species in question occupied a continuous and
unique area.

" 7. Hence an inquiry into the distribution of the individuals of a species in both time and space results in

the maintenance of the theory of the *unity of specific centres.*

" 8. Moreover, when we are able to trace the history of a species in time, we find, in the majority of instances, that there is a distinct indication of a paucity of individuals as we approach the epoch of its first appearance.

" 9. And when we are dealing with well-marked and continuous areas of species belonging to the present epoch, we find that there is within such area the indication of a point of maximum development of individuals, around which their numbers diminish.

" 10. We infer from these facts (8. and 9.) the probability of a single point of origin for every species within its centre of occupation.

" 11. In the course of time, however, it is possible that the area of occupation of a species may become removed from the point of origin, or may, after being removed, eventually return to its original position.

" 12. The indications of a single point of origin for each species, combined with the fact that we have no knowledge or experience of the individuals of any species being produced otherwise than from individuals of its own kind — in other words, that we have no knowledge of any other relationship between the individuals of a species than that of *descent* — leads to the inference that each species originated from a unique stock or prototype, consisting of a single being, or pair of beings, according as would be required for propagation.

" 13. Hence the point of origin within a specific centre is the point of appearance of the prototype.

" 14. How that prototype originated we know not ; but the doctrine of specific centres, originating each with a prototype of its own, is necessarily opposed to the hypothesis of *the evolution* of all species from one first form, without respect to the superiority or inferiority of the form.

" 15. That the prototype presented the *specific characters* (*i. e.* distinctive and constant features) of its descendants, is an hypothesis rendered probable by all that we know of the history of species in space and time.

" 16. The observation of the distribution of species in space and time indicates geographical areas and chronological epochs — points in time and space — where, as it were in preference, many species originated in groups. These we term *centres of creation*, and the phenomena of provinces are linked with the existence of them.

" 17. The value of Palæontology to Geology depends on the assumption of the constancy of specific types, and the unity of their centres or areas of occupation in time.

" 18. What is true with existing species should be *à priori* true with extinct ones, since we can clearly show that all known creatures, recent and fossil, are members of one biological system."

No. IX.

On the Recent Origin of Man.

Note to p. 495.

A discovery of fossil *human remains* has been recently made under circumstances which appear to me to call for much more close examination than (as far as I am aware) appears to have been bestowed on the case.

The statement to which I refer is given in a paper on

"The Railway Cuttings at Mickleton Tunnel," &c., by
G. E. Gavey, Esq., Quarterly Geological Journal, Feb-
ruary, 1853, p. 32., where the fact of the occurrence of
these remains is passed over without any comment.

The annexed sketch is taken from that accompanying
the paper, the *proportions* only being exaggerated for
clearness.

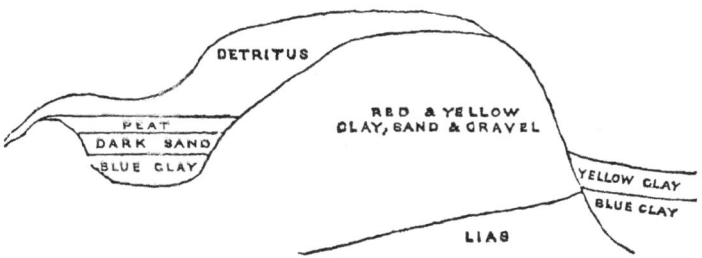

The material facts are briefly these : — The section of
the hill presents a basis of lias, immediately over which
is a mass of *yellow and red* sand, clay, and gravel, forming the summit ; on the northern side, after a mass of
superficial detritus, there was displayed what had been a
small hollow or depression in the slope of the hill, the
upper part of which was filled up by an ancient peat bog,
containing bones of various animals, probably of existing
species (but this is not distinctly specified); below this
occurred a bed of dark sandy earth, with stumps of trees
in situ; and below this a bed of indurated blue clay (the
same as that forming a continuous band on the opposite
face of the hill), in which a *human skeleton* was found,
in an inclined position, distinctly as *engulphed, not buried.*
Now on this I would observe : —

1. The beds composing the upper part of the hill con-

sist solely of *yellow* and *red* clays, and sands, &c. ; hence
the *blue* clay *could not* have come from the upper part of
the hill by washing down at a recent period, but could
only have originated in a deposition from the waters
before the hill had emerged to the level at which it occurs, and which is nearly the same on both sides.

[The *yellow* clay which covers it on the south *may
possibly* have been washed down, but may more probably have been a similar deposit.]

2. After the hill had emerged, the *blue clay* on the
south was dried up and consolidated. But on the north
the small patch of it continued exposed, and did not dry
up or consolidate, but remained in the state of *soft mud*,
probably by the retention of water in the hollow, at least
until the *human remains* were inclosed in it, which must
have taken place while it was in a soft state (unless,
indeed, they were engulphed while it was still under the
sea, which is not likely). *How long after the elevation*
it thus continued in the state of a muddy pond before
being filled up and coated over, is another point not
easily settled.

3. At some period *after* these remains were imbedded,
the water drained off, the clay consolidated, and a bed of
dark sandy earth overspread the top of it. Whence was
this derived ?

4. In this, *trees* then grew, their roots striking down
into the clay ; their stumps being found *in situ*, with
various other vegetable remains in the sand.

5. After the trees had fallen and decayed, a peat bog
formed on their site. All these events must have required a long series of ages. The peat bog, when at
length formed fully, must have remained such for a very

long period to account for the great *number* of remains of various animals — oxen, horses, deer, boars, and foxes — all collected in so small a space.

6. After this the peat dried up, and the whole became covered with a deposit of loam, sand, &c., which overspreads the side and top of the hill (but whose date is not explained), and which forms the modern surface and contour of the hill.

Considering the very long series of physical events which thus *must* have occurred *since the human remains were imbedded*, it becomes an important inquiry to endeavour to settle the probable relation of these various changes to any known epochs of geological action.

I merely wish to place these facts in a prominent light for the sake of exciting inquiry on the part of those better able to judge, and without pretending to offer an opinion on the point. The question would probably involve attention to the repeated series of changes of level and condition which have occurred in the long period since the pleiocene, especially as illustrated by the researches of Mr. Trimmer.*

No. X.

In connexion with what I have said † on the multiplication of species, some very important remarks have since been brought to my notice in Dr. J. D. Hooker's " Introductory Essay on the Flora of New Zealand."

---

* See Geol. Quart. Journal, No. 36. p. 293., and previous numbers there referred to.
† P. 397.

That distinguished naturalist contends that a great and needless arbitrary increase in the number of assigned species is continually being made by naturalists, from not sufficiently considering the actual very wide limits of departure and variation from any fixed or assumed type in each species. Thus the real number of species is much less than is often supposed (see § 2. pp. xiv—xvi.).

These statements, while they in one point of view would tend to modify any argument from the increasing number of species, yet in another would support the consideration that, even among existing species and within the limits of *finite* time, the power of change is so great as materially to impair the idea of any real principle of immutability, and thus to give greater scope to the possibility of more extensive modifications in the course of natural changes operating through *indefinitely extended periods* of past time.

The learned and acute author speaks of the possible " creation " of distinct and new forms in several places, but manifestly without restricting it to any particular hypothesis as to *mode* of production.

He regards the existing flora of the South Seas as the partial remains of that of more extended lands submerged, and speaks with disapproval of the hypothesis of the " creation of each species on each island by progressive developement on the spot " (p. xxi.), which, for those islands, may no doubt be perfectly just.

Yet again (p. xxv.), he disclaims entering on the question of the origin of species ; and while he considers them to have been permanent for ages, yet he observes there is nothing in what he advances inconsistent with *any* theory of their origin which the speculator may adopt.

After speaking of the obliteration of some species, he observes : " Whether the balance of nature is kept up by the consequent increase of the remainder in individuals, or by the sudden creation of new ones, does not appear nor have we any means of knowing." . . . . " We know that species perish, suddenly or gradually, without varying into other forms to take their place as species." (P. xxvi.)

These expressions are not perhaps designed to bear precisely on the present discussion ; but it may be remarked that the last sentence involves an *assumption* of the question here considered.

But it is alleged, in dwelling on the uncertainty of the limits of species, it is not meant that that uncertainty exists really in *the nature of things*, but only in the want of sufficient evidence and opportunities on the part of *naturalists*, that the indefinitude is not real, but only apparent—from our ignorance.

It must be recollected however, that the same argument will tell in the opposite sense, and that the same confessed ignorance should equally hinder the assertion of the immutable character of species.   Until the limits have been defined without fear of error, it is vain to assert that they are fixed.

Remarks of a similar kind are also advanced more recently, in the Introductory Essay to the " Flora Indica " by the same author in conjunction with Dr. Thomson (1855, p. 21.).

The authors speak of the hypothesis of " universal mutability " of species as opposed to facts, in which even the advocates for mutability, under special conditions in indefinitely long periods, would entirely agree.   They are also anxious to show, what would be equally conceded, that, even admitting that hypothesis, it would not invalidate systematic classification during the existing epoch.

The extensive limits of variation from a given type, in all species, is also particularly insisted on, and that, in fact, the true conception of a species lies *not* in any existing type or form, but in an *abstract ideal,* which an unpractised observer will often not recognise in familiar forms really belonging to the species. (P. 35.)

No. XI.

Note to p. 53.

In reference to my allusion to Comte's omission of geology in his " View of Positive Philosophy," an able and acute correspondent has suggested that the omission arose from the circumstance that Comte formed his system of science on the principle of a classification of laws, beginning with the most simple, as those of inertia, gravity, up to the most complex, those of life; whereas geology refers to a particular class of *phenomena,* not a peculiar set of *laws.* Supposing this to be the principle of his classification, it merely shows that that principle is defective for a complete or comprehensive system of science. But I am disposed to regard his fundamental idea as something different from this, and as referring rather to the perfect definiteness of the conceptions involved, and the exclusion of all hypothetical ideas.

No. XII.

Extracts from the Anniversary Address of Mr. W. J. Hamilton : 1855.

Quarterly Journal of Geological Society, No. 42.

" Thus, wherever we find the strata conformable, we have a confirmation of the well-known saying, ' Natura non facit saltum.' In fact all natural changes are gradual under these circumstances. The conditions of life gradually change; and the organic forms are modified to meet these changes. Certain species disappear, while others adapted to the altered circumstances are called into existence, and continue to flourish side by side with some of the pre-existing forms, thus confirming the view already stated, that when the strata are conformable, no line can be drawn between successive formations — the gradual change is not marked by sudden breaks in the series of animal life. In fact, we must not forget that our no-menclatures are for the most part only relative. Nature ever acts upon one long unbroken plan, and knows as little of sharp limits between Trias, Lias, and Jurassic, as be-tween the families and genera of existing organic life· These terms are at best but temporary shifts to assist our memories and enable us to register our facts and our knowledge.

" We must be careful not to give too much importance to nomenclatures, which deserve at the best but a secondary consideration." (P. lxviii.)

Again, in a subsequent passage the President expresses his view of the matter, more fully and generally, as follows : — We have found, during late years, that in pro-portion as we extended our knowledge of different forma-

tions, we have been compelled not only to introduce
a greater number of principal formations, but to sub-
divide these again into groups, and again to sub-divide the
groups into distinct beds. This process has long continued. We are no longer satisfied with primary,
secondary, and tertiary epochs ; it is not enough that we
have introduced the Permian, Neocomiac, and similar
terms, to designate different periods, or that we have sub-
divided the secondary rocks into Triassic, Liassic, Jurassic,
and Cretaceous ; all these divisions are again sub-divided,
I might almost say, *ad infinitum.* As the investigation of geologists has extended itself over distant
countries, and has brought fresh continents under our
notice, new, and at first sight anomalous, combinations
have been brought to light. The limits and breaks
already assigned to different formations, in the countries
where first observed, have not been found always to hold
good. The marked unconformability of stratification, and
the distinct differences of palæontological evidence, on
which the limits of formations were first grounded, have
in other countries either disappeared altogether, or have
required to be greatly modified. It has been found that,
between these respective limits, as at first laid down, certain fossils of the lower beds extend higher up into those
above, while some of those hitherto supposed to be characteristic of the overlying formation are found extending
downwards into beds of an older age. On the other
hand, that unconformability of strata which was supposed
to mark the limits of epochs, and to point out the breaks
occasioned in the successive deposition of strata by great
natural convulsions, is often found to disappear when the
investigation is extended and the strata are traced into

other countries. In this dilemma, the first step has been to intercalate new beds, as intermediate between the different formations, connecting them as it were by a certain community of animal life, marking the passage from one condition of existence to another; as, for instance, the S. Casciano beds are now introduced between the Triassic and the Liassic, the Carboniferous Shales between the Old Red Sandstone and the true Carboniferous beds, and others, which will readily occur to you. But the difficulty does not cease here. As we extend our inquiries, we find that the gradual passages from one formation to another are mere local phenomena; and we are thus almost forced to the conclusion that such marked separations between the different formations, as we have been fondly trusting to, do not really exist in nature. I believe the time will come when, having brought before us a greater amount of sections all over the world (if, indeed, it is not possible to do so already), we shall find that there exists a gradual passage from the very oldest to the newest strata, that from the earliest fossiliferous rocks to the most recent post-pleiocene formations, there has been one unbroken sequence of deposits, modified only by local disturbances, showing the gradual change of organic life according to the different conditions of existence; that in every case a certain number of species existing in the beds below have been continued upwards, mingled with new forms specially created to suit the new state of things; and that this progress has ever been going on in some part of the earth's surface, undisturbed by other local changes and convulsions. We know that as the conditions of life varied, new forms we recalled into existence, while former ones were gradually disappearing ; but we shall, I think,

be more and more forced to give up that view which led
us to subdivide the countless myriads of ages of geologic
time into epochs, formations, groups, and subdivisions,
and to look upon the whole series as one grand group,
modified in time by a slow and imperceptible progress,
and affording breaks and interruptions of conformability
of strata only as local phenomena." (P. xci.)

No. XIII.

On the Argument of Natural Theology.

In what has been advanced, in several parts of each of
these Essays, on the subject of *Natural Theology*, I have
adverted very little to the *metaphysical* or *moral* proofs
of the existence or attributes of the Deity, but have
confined my remarks entirely to the *physical* evidence,
and the strict conclusion from it. It may be desirable
to add a word in reference to those other modes of
reasoning, especially in connexion with what I have remarked in a previous part of this Appendix as to the
relation of my argument to metaphysical views.

In the present instance such a reference to the metaphysical speculations of some of the most eminent philosophers, especially those of KANT, afford a strong
corroboration of the propriety of the course I have
pursued. It is perfectly well known that the various
alleged *à priori* proofs of a Deity (as those of Descartes,
Leibnitz, Clarke, and Locke), after being analysed in a
masterly manner by that great metaphysician, have been,
as I think, conclusively shown to fail as strict philosophical arguments.

He, however, fully admits the physical argument, but

as appears to me, without giving it that primary importance which I conceive it deserves.

He dwells much more upon the moral and practical argument arising from the common feeling of mankind; and doubtless, in a *practical* point of view, nothing is more powerful than such an appeal; but this is confessidely *beside the question* of strict philosophical evidence.

Again, abstract and *à priori* arguments, if ever so valid, can lead to nothing but abstract conclusions; the idea of a Deity, so deduced, can be nothing but a mere abstraction and creation of the intellect. This, then, would bring us very nearly to the theory of Feuerbach, — "God exists only in our minds." Thus it appears to me more especially desirable to dwell on the *physical* argument, and particularly important to be careful to present it in its strict and legitimate form, as to the *kind of conception* which it furnishes. Granting whatever force may legitimately belong to any of the other arguments alluded to, it is clear that the *physical* evidence is precisely that which is the only real *corrective* and *corroborative* of them all.

In the speculations of Feuerbach here alluded to, which have obtained much celebrity (Essence of Christianity, transl., London, 1854), there is also much bearing on some other points discussed in these Essays.

This is not the place to go into any observations on his theory of religion in general. I shall merely advert to one or two particular points having reference to the relations of theology to science.

One of Feuerbach's leading ideas (as a consequence from his principle) is the essential antagonism between

the religious principle and the contemplation of external
nature. Hence the desire to render the universe subordinate
to man — hence the hostility to " a philosophy of second
causes," and the idea that " religion is abolished when
second causes are interposed between God and man.'
(Transl., p. 180.)

He refers especially to the idea of " creation," as essential to the religious idea, and remarks on its peculiar
significance in the Judaical system (p. 112.). Yet it would
seem as if he confounded this with the Christian doctrine,
as indeed the mistaken views of many divines might
justify him in doing.

He dwells, with much force, on the false and narrow
philosophy of a class of writers who dwell solely on the
low utilitarian view of nature; he remarks the philosophical inconsistencies of their advocacy of " creation "
(pp. 84. 190.), as well as of their contracted view of final
causes, though he is not particularly happy in the
instances which he selects (p. 103.).

In these remarks (his translator informs us) he had an
eye to " the vapid and narrow theology of the English
natural philosophers ;" and his censures, I conceive, are
not wholly undeserved, though I venture to hope the
views advanced in my Essays may be admitted as an exception to this sweeping charge.

Whatever may be thought of Feuerbach's speculations
as a theory of real Christianity, they certainly evince a
deep insight into the working of the tendencies of human
nature towards those corruptions and excesses which too
often usurp the name of Christianity, with which he
seems to confound it, and the study of which fully explains,
on a common principle, the antagonism between that fana-

tical spirit in its diversified forms and all philosophical views. To those views, even in themselves, the vulgar mind feels a natural antipathy ; and when to these the demands of superstition are added, we have a ready clue to all the delusions, extravagances and incoherences popularly broached on such subjects, which are but the expression of a religious animosity against whatever tends to humiliate man's imagined self-importance, some instances of which have been adverted to in these Essays.

Hence we may understand the pious horror with which all new discoveries and applications of the powers of nature are regarded ; hence the sacred jealousy of inhabitants in other planets ; hence the profaneness of the nebular hypothesis, " the dull and dangerous heresy of the age ; " * hence the still more flagrant wickedness of the theory of development, and the high merit of those scientific men who pander to the popular religious appetite by denouncing such views ; hence the sin of geology, and the righteousness of those who seek to do away the offence even by the most transparent subterfuges and evasive compromises.

The same spirit descends, on the one hand, to dictate a religious faith in the existence of live toads immured in solid rock from the creation, or full grown animals brought forth out of the earth ; on the other, soars to the assurance that the whole universe is merely subservient to the supreme dignity and importance of man — the planets created only to be the locality of his future existence—the commencement of *his* species the only epoch

* Brewster, " Life of Newton," ii. 31.

worthy the name of creation — the earth, as his abode, the *moral* centre of the universe, while its position as the *physical* centre is but reluctantly denied, nay, may be even still open to question. The rotation of the moon on its axis is authoritatively condemned! and that of the earth itself rests on arguments little better! Foucault's experiment (so eagerly grasped at by the Copernicans) has been explained on quite other principles!

We are thus in all points veering fast towards the old and orthodox Ptolemaic doctrine, which will, doubtless, soon be stamped with the imprimatur of the Inspectors, and taught in our national schools, along with the creation of the world in six days, as indisputable Scripture truth, and all impugners of either handed over to the ecclesiastical tribunals.

THE END.

ERRATA.

Page 160., line 2., for "then" read "there"

" 302., line 3., for "Can" read "Or can"

" 348., line 2., for "animal" read "organic"

" 471. *note,* for "Religious" read "Religion"

" 486. *note,* for "profane" read "as profane."